WOOL

Unraveling an American Story of Artisans and Innovation

Peggy Hart

Schiffer Publishing Ltd
4880 Lower Valley Road • Atglen, PA 19310

Copyright © 2017 by Peggy Hart

Library of Congress Control Number: 2017935614

All rights reserved. No part of this work may be reproduced or used in any form or by any means—graphic, electronic, or mechanical, including photocopying or information storage and retrieval systems—without written permission from the publisher.

The scanning, uploading, and distribution of this book or any part thereof via the Internet or any other means without the permission of the publisher is illegal and punishable by law. Please purchase only authorized editions and do not participate in or encourage the electronic piracy of copyrighted materials.

"Schiffer," "Schiffer Publishing, Ltd.," and the pen and inkwell logo are registered trademarks of Schiffer Publishing, Ltd.

Designed by Danielle D. Farmer
Cover design by RoS
Type set in Alexon / Freehand575 BT
Front cover and title page image: Sheep on the White House lawn, 1918. Library of Congress.

ISBN: 978-0-7643-5431-1
Printed in China

Published by Schiffer Publishing, Ltd.
4880 Lower Valley Road
Atglen, PA 19310
Phone: (610) 593-1777; Fax: (610) 593-2002
E-mail: Info@schifferbooks.com
Web: www.schifferbooks.com

For our complete selection of fine books on this and related subjects, please visit our website at www.schifferbooks.com. You may also write for a free catalog.

Schiffer Publishing's titles are available at special discounts for bulk purchases for sales promotions or premiums. Special editions, including personalized covers, corporate imprints, and excerpts, can be created in large quantities for special needs. For more information, contact the publisher.

We are always looking for people to write books on new and related subjects. If you have an idea for a book, please contact us at proposals@schifferbooks.com.

Other Schiffer Books on Related Subjects:

Dyeing Wool: 20 Techniques, Beginner to Advanced, Karen Schellinger, ISBN 978-0-7643-3432-0

Felting, Elvira López Del Prado Rivas, ISBN 978-0-7643-4531-9

The Art of Weaving a Life: A Framework to Expand and Strengthen Your Personal Vision, Susan Barrett Merrill, ISBN 978-0-7643-5264-5

Contents

INTRODUCTION: *Why Wool?* 5
COLONIAL PERIOD *1609–1780* 17
NEW NATION *1780–1840* 41
INDUSTRIALIZATION AND MECHANICAL INNOVATIONS
    1840–1890 71
GOLDEN AGE OF INDUSTRY *1890–1920* 91
WOOL'S GAINS AND LOSSES: *American Cultural Trends*
    1920–1950 113
LOSING GROUND: *Post–World War II 1950–1980* 132
EPILOGUE: *Wool: Necessity to Niche* 157

ACKNOWLEDGMENTS 172
NOTES 174
WOOL (MOSTLY) GLOSSARY 182
BIBLIOGRAPHY 185
INDEX 191

INTRODUCTION
Why Wool?

"The history of wool and its bearer, the humble but persistent sheep, parallels that of man. Together they emerge from the shadows of unrecorded time into the light of known history, each jointly dependent on the other."

Romantic Story of Man and Sheep,
Pendleton Woolen Mills

Shepherd and sheep sculpture, clay.
Photo by Joanne Semanie

The story of wool in the United States is a romantic yarn, a story of love and love lost. It is an epic tale with a cast of millions: farmers, farmwives, shepherds, sheepherders, artisans, merchants, immigrants, slaves, Native Americans, inventors, captains of industry, millworkers, advertising and sales people, department store owners, husbands, and housewives. The backdrop of American history is the set: colonization, slavery, wars, the Industrial Revolution, westward expansion, cultural shifts as brought about by innovations such as central heating and automobiles.

The plot follows the sheep, source of the original miracle fiber, as they arrived in the New World and spread from coast to coast. For the first three hundred years, wool was grown, processed, consumed, and cherished as one of life's essentials. Then, in a sudden plot twist akin to an alien invasion, it met a fearsome foe, man-made fibers, which won the hearts of Americans with new garments like acrylic sweaters, men's polyester double knit suits, and finally polar fleece jackets.

Wool's fade to near irrelevancy by the end of the twentieth century was drastic and unforeseen.

As late as 1940 a monograph written about the woolen and worsted industry in Rhode Island stated "the importance (of wool products) in everyday living is obvious; so important that, since history has been recorded, they have been taken for granted and always will be."

The 1950s proved to be the pivotal point, the beginning of the end for the commercial American wool industry. Post–World War II, man-made fibers became widely available to the textile industry, weaving technology evolved further, and consumers began to make different choices. Dr. Ernest Dichter, president of the Institute for Motivational Research (of consumers), asked in an *American Fabrics* article in 1958, "How does it happen that a centuries-old romance between man and textiles has lost so much of its ardor? That an earlier intimacy and understanding has given way to cool aloofness?"

The story of wool traces the arc of production from handcraft to mechanical invention and industrialization, returning to handcraft. Though deindustrialization in the twentieth century closed almost all the mills, handcraft would keep the noble fiber alive, a development that is celebrated at the sheep and wool festivals held in many states.

Seasonal grazing in Horse Valley, Dixie National Forest, Utah, 2016.
PHOTO BY DAN WARNER

Wool, the Original Miracle Fiber

Wool is an animal fiber produced by sheep (sheep can have hair or wool), and varies widely in length, fineness, strength, color, shrinking tendency, and luster. Wool's physical properties vary greatly from breed to breed, sheep to sheep, even belly to back.

Sheep breeds produce fiber ranging from very fine to very coarse. The speed of growth can range from less than an inch to 12 inches each year, though for most breeds somewhere between 3 and 5 inches is typical. Fine fiber, most desirable for clothing, is produced by Merino and related breeds. Coarse, long staple wool, coming from breeds such as Karakul, is durable in carpets.

It is thought that sheep were one of the first domesticated animals, dating from 9000 BCE. The first evidence of wool fabric dates to around 2500 BCE, with records of its use since the Roman Empire. In 38 BCE, 200,000 sheep were bequeathed to the Roman emporer Augustus by a single devoted citizen. It's likely these sheep were Tarentine, a now extinct breed known for its very fine wool. It was so precious that the sheep wore coats to protect the fleece, a practice still employed today by some shepherds. The Roman poet Virgil composed his Eclogues during this time. An eclogue is defined as a short poem, especially a pastoral dialogue. The voice in the poems is that of a shepherd, celebrating many things during his idle hours, among them the glory of the pastoral life.

Since "the light of known history" mentioned at the start of this chapter, people continue to keep sheep, customizing wool by crossbreeding to achieve fibers of different lengths, diameters, and degrees of softness to accommodate the changing requirements for clothing, carpets, and other household items.

Wonders of wool.
COURTESY OF AMERICAN SHEEP INDUSTRY

Perhaps wool's most unique feature is its structure, a shaft with scales. These scales allow wool to trap heat, breathe, shed water, and stay warm when wet; this is how it earned its title as the original miracle fiber. It is inflammable, which is why you can use a wool blanket to put out a fire (of course, wet blankets work even better). It is infinitely plastic because it is the only natural fiber having an innate crimp. This crimp gives the wool fiber resilience, allowing it to be stretched up to 50 percent of its length, and then return to its original dimensions.

A story from the *Textile Industries Magazine,* April 1964, illustrates the miracle of wool:

> WOOL WET BUT USABLE. A cargo of South African wool which has been lying in about 160 feet of water 50 miles off Nantucket Island, MA, for 22 years will be salvaged and used, a Boston firm reports. The wool—14,000 bales of it—is in the freighter *Oregon,* which sank after a collision during World War II. Divers report that the wool, worth $5.5 million, is in good condition except for a quantity near the point of impact, which is saturated with bunker oil.

Populations—Natives and Immigrants, Ovine and Human

Like the American people, sheep are largely immigrants. Two species of wild sheep are native to North America, bighorn and Dall sheep, both of which are thought to have crossed into North America from Siberia on the Bering land bridge. Bighorn sheep range from southern Canada to Mexico, and have fur like deer rather than wool. Dall sheep have a fine wool undercoat, befitting their arctic and subartic habitat of Alaska, Yukon, and British Columbia.

Bighorn sheep.

The first European sheep, Spanish Churro, accompanied Coronado on his 1540 exploratory quest for cities of gold in the North American southwest. The next transatlantic immigration wouldn't be until English settlers brought English sheep, probably Wiltshire, to Jamestown in what would become Virginia. Assorted English and Dutch breeds followed intermittently and then in 1811 came the Merinos, another Spanish breed. Sheep would accompany the settlers as they moved west, with the nature of sheep raising changing from farm flocks to range animals, until most American wool was produced in the western states. From the earliest history of sheep raising in America, American farmers never produced enough wool for the nation to be self-sufficient. However, the history of sheep raising, and the number of sheep in the US over the centuries, does go hand in (wool) glove with manufacture and consumption, as we'll see.

The American people as producers and consumers of wool are key to the story. Who were they? Where were they? Native Americans, immigrants from all parts of Europe, and the involuntary immigrants from Africa all were producers and consumers of wool in the United States. Like the sheep themselves, they are of various origins and on the move, forever in search of new grazing. The story of wool necessarily includes detail about population, immigration, slavery, and migration from rural to urban areas.

Early information about wool can be gleaned by listening to the voices of colonial settlers, slaves, senators, and sheepherders, as documented in places like diaries, slave narratives, accounts of life on the range, and arguments in eighteenth- and nineteenth-century legislatures. Later sources of information about wool include government publications such as the Census of Manufactures beginning in 1810, the United States Department of Agriculture (USDA) surveys, industry periodicals (like *American Fabrics*) and consumer periodicals (like *Delineator* and *Ladies Home Journal*), manufacturers' sample books, and company annual reports.

Technology, Industry, and Politics

In colonial America, wool textiles were manufactured entirely by hand, by either knitting or weaving them. However, soon after the founding of the country, textiles would begin to be made with machines. Along with

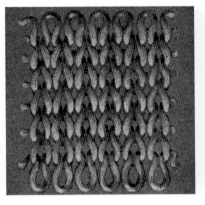 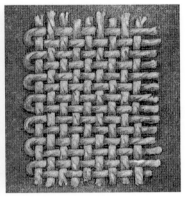

LEFT:
Knit cloth construction, from *Fabrics and How to Know Them*, 1928.

RIGHT:
Plain weave cloth construction, from *Fabrics and How to Know Them*.

cotton, the wool industry was at the forefront of the American Industrial Revolution, beginning around 1760. It began in New England because New England was already using water power, and had raw materials. It also had inventors, millworkers, millwrights, and captains of industry.

The wool industry lurched along through the cataclysmic events of westward expansion, wars, immigration, the Depression, and changes in lifestyle. As wool was the fabric specified for provisioning soldiers (uniforms, overcoats, and blankets), each successive war stimulated and supported its manufacturing. Many cotton mills converted to wool during the Civil War and the industry doubled between 1859 and 1869. In 1864, 60 million pounds of wool were produced for the army alone. Uniforms for workingmen of all sorts were the first ready-made garments, in fact, and the invention of the sewing machine in the 1850s helped to accomplish outfitting the armies. But bunting fabric, the fabric used in flags, came from England even as late as 1866. This was a source of chagrin for politicians, who enacted the first important tariff for wool in 1867. The Wool and Woolens Act of 1867 levied a duty on imported woolens, and was followed by the McKinley, the Wilson, and the Dingley Tariffs. Tariffs helped ensure that the military, relying heavily on wool for uniforms and blankets, would have an adequate domestic supply.

The American wool industry enjoyed a golden period from 1890 to 1920. Demand for wool apparel followed exponential growth of the US population, from 5 million in 1800 to 76 million by the end of the century. Cloth was in high demand.

The Crompton & Knowles Loom Works of Worcester, Massachusetts, was especially inventive as the primary manufacturers and suppliers of looms to the woolen and worsted industry during this period. They invented the automatic fancy loom, truly one of the most complex and ingenious machines to come out of the Industrial Revolution. The company listened to the suggestions it received from managers at mills, and established one of the first research-and-development departments to make innovations. In the hundred years between the first patent in 1837 and 1937, Crompton & Knowles acquired more than 2,100 patents.

New fibers stimulated the next period of technological innovation in fabric construction, beginning with the development of rayon in the 1920s. From the 1920s to the 1960s the American textile industry reinvented itself. Yarn was engineered rather than spun, made from synthetics rather than natural fibers, and textiles were constructed using processes that often had little to do with weaving. The industry evolved by developing new fibers, new fabrics engineered with the new fibers, and even new textile categories such as bonded fabrics and techno textiles, textiles engineered for a multitude of purposes, very few of which fit the traditional concept of textiles. Weaving technology changed radically, with the invention of shuttleless looms that could run at faster, more profitable speeds. What did this mean for the Crompton & Knowles Loom Works? Until 1977, Crompton & Knowles was still building shuttle looms primarily for weaving natural fibers. Their earlier innovative practices had lapsed and the Loom Works discontinued production of looms, at the same time that most of the New England woolen mills were closing.

In the early 1900s, there were 800 woolen and worsted mills operating in the US. These mills, located mostly between Pennsylvania and Maine, produced a variety of fabrics, from suiting to car upholstery to blankets. Filled with the rattle and bang of shuttles or the whir of spindles, they employed hundreds of thousands of people and represented the seventh largest industry in the country in 1939. The buildings began emptying of their workers in the 1950s. Eventually the buildings were also emptied of the machines, which were largely sold for scrap, although occasionally the spinning frames and looms made their way across oceans to live new lives in developing countries. Today, skeletal remains are all that is left of once thriving manufacturing cities such as Woonsocket, Rhode Island, and Fall River, Worcester, and North Adams, Massachusetts.

Consumption

In the early 1700s, near the beginning of this story, wool use was ubiquitous in America: colonists wore short gowns and long gowns, petticoats (skirts), smocks, breeches, and waistcoats, cloaks, greatcoats, and stockings made of wool. They slept under wool blankets, and kept themselves warm in carriages under wool carriage robes. Even as recently as a hundred years ago, wool was still found in every part of the wardrobe, from underwear to overcoats. Wool flannel was a staple for things like shirts, underwear, and petticoats. The original tennis whites were wool, as were bathing suits and army uniforms. Everyone knows about silver and gold wedding anniversaries, but a gift of wool may be given for the seventh anniversary (as an irreverent friend quipped, "the seven-year itch"!). The consumption of wool per capita per year in the US was 4.49 pounds in 1840. It increased steadily, from 8.52 pounds in 1880 to its high point of 9.07 pounds in 1890. But per capita consumption today is 1.5 pounds, the equivalent of two pairs of socks.

Wool's decline in the one hundred years following 1890 is a story of technological and social change, marketing forces, and above all, consumer choices. Paul Cherington, marketing research pioneer and member of the original faculty of the Harvard Business School, wrote in his 1932 book *The Commercial Problems of the Woolen and Worsted Industries* that until 1920 what the mill produced the buyer accepted, but after 1920 the market changed rapidly. Heated trains and steam-heated apartments and houses eliminated the need for heavy woolens; people were wearing fabrics half the weight of what they had previously worn.

Lewis Miller, watercolor chronicler of daily life, showing short gown, petticoat, and long gown. 1805.
FROM THE COLLECTION OF THE YORK COUNTY HERITAGE TRUST, YORK, PENNSYLVANIA

Consumers could choose from a wide array of fiber options. Synthetic fibers were first introduced to the public with the exhibition of "artificial silk" (later christened rayon) fibers at the International Exhibition in Paris in 1887. In 1911, a factory in Pennsylvania produced 362,000 pounds of the new fiber. By 1926, American factories were producing 63 million pounds. The name *rayon* was copyrighted, and eventually entered the dictionary as a word in the same category as cotton, wool, silk, linen, and jute. Subsequent man-made fibers followed: nylon, acetate, Dacron, Orlon, Acrilan, and most recently carbon fiber and Kevlar. Though wool had faced some competition from cotton and linen from the beginning of the European settlement of North America, the natural fibers were used for different purposes, and generally coexisted peacefully. This was the first time wool faced competition from another fiber that claimed to do what wool did best: supply warmth.

The availability of new fibers and fabrics, and the changes in how people lived, resulted in changes in consumer fabric preferences. Wool's disadvantages came to outweigh its advantages around 1950. From 1950 on, wool use diminished in apparel, car upholstery, and home furnishings. By 1969, per capita consumption had dropped to 1.5 pounds, not budging for fifty years.

Today, wool invariably evokes a visceral response; usually either enchantment or revulsion. People who are not fans of wool often claim allergies, or the difficulty of caring for wool articles. Polyester fleece has largely supplanted wool for these folks. American people have largely lost their wool literacy, that is, the appreciation of wool's characteristics and knowledge of how to care for it.

Early settlers might not have worn cotton, but they all wore wool. The reverse is true now: all Americans wear cotton, but many may never wear wool.

Wool in Context

My telling of the story of wool in the United States looks at wool within the context of American history and culture. It is a history of the consumption of a commodity as much as the manufacture of it. What did consumers take from what was offered, and why? Consumption includes the satisfying of basic needs, and also the gratification of wants and desires. Consumer needs shaped the evolution of wool items such as blankets, broadcloth cloaks, overcoats, and the ubiquitous blue serge suit. The wool industry was impacted by cultural trends, and its fate depended on how it responded.

Comprehensive accounts of sheep raising and the wool industry include Cole's *American Wool Manufacture,* published in 1926, and Edward Wentworth's *America's Sheep Trails,* published in 1948. It's obvious that the story of wool has evolved quite a bit since those books were written. My aim is not to write the business or economic history, but to tell the story from the vantage of a craft practitioner, someone who came to weaving using industrial equipment at exactly the moment that all the mills were closing. In 1981, I worked as a weaver in the samples department at one of the last mills in Rhode Island, the Worcester Textile Company in North Providence. In 1982, I bought my first Crompton & Knowles loom, and began producing blankets by machine, after years of handspinning, plant dyeing, and handweaving. These 1940s vintage looms are great; they allow me to weave thousands of blankets a year. I concur with Ernest Batchelder, turn of the century American designer, who said "if the medieval craftsmen could return to the world, they would welcome machinery."

Having a foot in both the handcraft and industrial machine world helps me appreciate handicraft while simultaneously enjoying the technological advances made with the mechanization of textile production. A disclaimer: I am an unapologetic advocate of wool, with the belief that even if one chooses not to consume it, its history matters.

Recording the history of wool allows us to remember textile fabrics, many now vanished: amazon, baize, batiste, Bedford cord, brilliantine, broadcloth, camblet, cassimere, challis, chinchilla, covert, delaine, domett, Florentine, gabardine, grenadine, Henrietta, hopsacking, kersey, Melrose, melton, negro cloth, poplin, prunella, satinet, serge, shalloon, unions, wadmal, whipcord, woolsey, zibilene (a heavy coating fabric with long shaggy nap laid in one direction; see glossary for other definitions). The functions and constructions of these are a window into other cultures, where a layer (or two or three) of wool made the difference between fashion or frumpage, comfort or misery, survival or extinction.

COLONIAL PERIOD
1609–1780

> "ALL THE WORLD WEARS IT, ALL THE WORLD DESIRES IT, AND ALL THE WORLD ENVIES US THE GLORY AND ADVANTAGE OF IT."
>
> Daniel Defoe, *Plan of the English Commerce*, 1730

To describe the fledgling seventeenth-century American wool industry one has to reference the British wool industry of the sixteenth and seventeenth centuries. From the fourteenth century on, woolen cloth was England's dominant manufactured product, its most valuable export, and so vital to financing the government that the Lord Chancellor, head of the House of Lords, was symbolically seated on a wool sack.

Wool use in clothing was a ubiquitous feature of everyday life in seventeenth-century England. Gregory King (1648–1712), an early statistician from preindustrial England whose major work, *Observations and Calculations, Natural and Political, upon the State and Condition of England*, was published almost a century after his death, filled dozens of notebooks with calculations and estimates, most of which remain unpublished. In the 51st notebook from 1688, now known as the Burns Journal, he was engaged in calculating "Annual Consumption of Apparel" and concluded that about half of all clothing in the seventeenth century period that he studied was made of wool. At that time cotton was imported from India but not yet widely available. From 1700 to 1774 the wearing of it was even prohibited in England. Silk was precious, coming from the even further away Far East. The only other available fiber for clothing was flax, from which linen is spun.

England's climate demanded warmth in clothing, so that meant wool, and the choice mainly was in the quality of wool used. Farmers raised sheep of various breeds in various locations, for various purposes, just as they do now. There were longwool breeds such as Romney and Leicester, used for making worsted fabrics. Medium grade wool included the East Anglia breeds and this was used for clothing, particularly broadcloth. Anything finer than Shetland wool had not yet made its way to England; Merino sheep were jealously guarded in Spain. Linsey-woolsey was the original union cloth, woolen spun wool woven on a linen warp. However, England did best with broadcloth. Broadcloth was its chief product for 500 years, its production being a series of highly specialized operations.

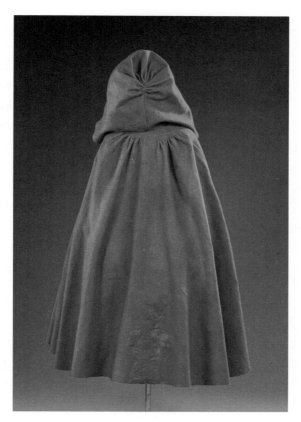

Hooded cloak, red broadcloth, circa 1785–1795.
PHOTOGRAPH COURTESY OF POCUMTUCK VALLEY MEMORIAL ASSOCIATION, MEMORIAL HALL MUSEUM, DEERFIELD, MASSACHUSETTS

What was broadcloth? We hardly use it now, but it was a plain woven fabric so heavily fulled, on its way to being felted, that it didn't need to be hemmed but merely cut to shape. The Pocumtuck Valley Memorial Hall Museum in Deerfield, Massachusetts, has a splendid example of a woman's hooded cloak. The fabric appears to be ¼" thick. Coarse broadcloth was used for outer garments such as coats, cloaks, and capes, and finer grades were used for uniforms and suit coats. Width on the loom could be anywhere from 85 to 95 inches (hence the name broadcloth; ordinary cloth was usually not more than 40 inches wide). Weaving on such a wide loom required two people, until 1733, when John Kay invented the fly shuttle. It both accelerated the process of handweaving and made possible the weaving of broadcloth with only one person. After weaving, the broadcloth was then fulled, a washing process deliberately used to shrink the fabric to a finished width of 54 to 63 inches. After fulling, the cloth was brushed to bring up a nap, and then sheared so that it was evenly fuzzy. Each step in cloth production was performed by a different guild of trained artisans, who memorialized their lost professions in their surnames: Weaver, Fuller, Sherman (shearman).

Populations, Human and Ovine

The Spanish were the first European explorers to reach the New World, and the first to bring domestic livestock. Hernando Cortes was ranching in New Spain (Mexico) by the mid-1550s, and Spanish settlement advanced northward into what would become the American Southwest.

Europeans began to migrate to eastern colonies in 1607, settling in Jamestown in what would become Virginia. The first sheep arrived in Jamestown in 1609, probably Wiltshires, a popular breed in England in the seventeenth and eighteenth centuries. Wiltshires are a hardy hair sheep kept primarily for meat, with a wooly undercoat that sheds in the spring. Early settlers may have engaged in woolgathering by collecting the fleece caught in weeds and fences, but wouldn't have collected sufficient amounts to make more than a sock. Later shipments may have included Leicestershires, a long wool breed. By 1649, there were said to be 3,000 sheep in Virginia, not a great number.

The Dutch settled the New Netherland (later New York) as early as 1625, bringing Texel sheep from Holland.

Pilgrims aboard the *Mayflower*, planning to join the Virginia company, were blown off course to the north by a storm and landed at Plimouth,

Massachusetts, in 1620 without sheep or other large animals. This was remedied sometime thereafter with Wiltshire sheep present by 1624. Probate estate inventories made in 1633 for three *Mayflower* passengers, Samuel Fuller, Francis Eaton, and Peter Browne, show that all three men owned several rams, sheep, and lambs. In about 1635, forty Texel sheep were bought by Pilgrims from Dutch settlers of Manhattan Island.

From 1629 to 1640, as many as 21,000 people emigrated from the East Anglian counties of England (Norfolk, Suffolk, Cambridgeshire, Essex, Lincolnshire) to New England colonies. During this period, people emigrated to New England in "companies," coalitions of family groups; there were also clerical companies, or groups of people who emigrated with a particular clergyman. This concentrated migration pattern fostered closer social cohesion, and offered greater support for the emigrants.

East Anglia was renowned as a cloth producing area, especially of broadcloth. In the mid-seventeenth century, the General Court of Massachusetts actively solicited sheep importation: "those having friends in England desiring to come, would write them to bring as many sheep as convenient with them, which being carefully endeavoured, we leave the success to God." It is reasonable to assume that settlers brought sheep, anticipating basic survival needs. Sheep supplied both wool and meat, and their droppings fertilized the soil. In the American Historical Society Report on the sheep industry, the author writes that the sheep were likely Wiltshires and Romney Marsh. He estimates that by 1642 there were 1,000 sheep and 10,000 people in Massachusetts. One sheep per ten people would hardly have been enough to supply wool clothing needs. At this time, however, arriving ships' holds were full of imported wool cloth and other necessities. By 1651 Captain Edward Johnson reported in his narrative *Wonder-Working Providence of Sions Saviour* that the number of sheep in Massachusetts had increased to 3,000.

The islands off of Massachusetts and Rhode Island, including Nantucket and Martha's Vineyard and those in Narragansett Bay, supplied an important environment for sheep raising. Sheep had a hard time getting established on the mainland of the New World, where wolves were numerous. Sheep needed to be placed under the care of herders, and pastured on town commons. Islands offered protection. Martha's Vineyard, a large island off Cape Cod, was settled by Europeans in the 1640s. In 1764 Hector St. John Crevecoeur visited and reported in *Letters from an American Farmer* a population of 20,000 sheep, a fabulous number relative to the 4,000 inhabitants. He also noted the "excellent pastures." Alas, the Vineyard's pastoral success would attract the attention of British Major General Charles Grey in 1778, when he was tasked with "procuring the necessary refreshments" for his troops during the Revolutionary War. Martha's Vineyard suffered the loss of 10,574 sheep in Grey's Raid. General Grey arrived with eleven men of war and other supporting vessels. The islanders professed willingness to comply with the demand, with the promise of payment. It took several days to collect the sheep and bring them to the dock, after which twenty vessels were filled with 6,000 sheep. Vineyarders submitted a petition for payment, which was ignored by the British government. Fortunately, not all the sheep were found at the time of rounding up, and the population rebounded sufficiently that a visitor, Rev. James Freeman, reported 15,000 sheep in 1807. Wool production was 23,400 pounds.

In Rhode Island, farmers successfully introduced sheep as early as 1636. Rhode Island, being composed of a number of islands, also offered excellent protection from wolves. By 1661, the human population stood at about 2,000, with 10,000 sheep. This was more than enough for the population's needs, and surplus sheep were exported to neighboring states.

Although the first sheep in Virginia were in the Jamestown settlement, the Virginia barrier islands created another protected environment for sheep. Smith Island was first used as a grazing meadow, offering the advantage of not having to clear land or erect fences.

In 1707, Scotland and England combined to become Great Britain, giving Scottish people legal access to emigrate to the American colonies. In 1718 a group of Scots arrived in five ships of clerical groups with their Presbyterian ministers, settling in Londonderry, New Hampshire. By 1772 there was a population of 30,000, of whom 10,000 were reportedly weavers. Later Scotch Irish and Irish groups would settle up and down the east coast.

WOOL
Colonial Period 1609–1780

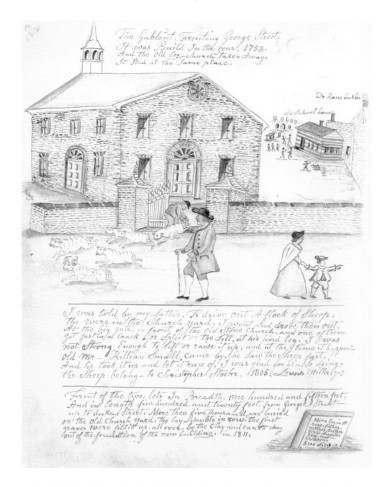

Lewis Miller, Pennsylvania, 1805, showing sheep and costume of the day.
FROM THE COLLECTION OF THE YORK COUNTY HERITAGE TRUST, YORK, PENNSYLVANIA

A diverse group of European settlers including Quakers, Germans, and Scots began to arrive in Pennsylvania in the late seventeenth century. Many were displaced cloth workers, who brought knowledge of the construction of specialized fabrics. In Pennsylvania, more than half of all households had sheep, though not necessarily kept for their wool.

All the European settlers, whatever their country of origin, had worn wool in their native land. Many immigrants to the American colonies possessed cloth making skills and passed these along to subsequent generations, sometimes formally through apprenticeships, or informally within families. The exception to this were the involuntary emigrants, enslaved Africans. But even slaves would be clothed in several types of wool fabrics (known as negro cloth) and taught skills to construct cloth designed specifically for their use.

The Politics of Clothing

The first European settlers often arrived with literally the clothes on their backs, and in the early years depended heavily on being provisioned by ships coming from England. Colonists wore what they were familiar with in the countries they came from, primarily wool and linen clothing. (At the time, linen was used for the same purposes that cotton would be later: garments and household textiles such as sheets and tablecloths.) Certainly early settlers wore a lot of wool. Examples of what they received from England can be seen in the records of the contents of the holds of ships entering Boston Harbor. The Aspinwall Notarial Records of 1645–1651 list:

> WOOLENS: cottons [not really cotton, but a type of wool fabric], Spanish cloths [not all woolens were produced in England], northern Kerseys, frieze, Leyden duffles, broadcloth, Kentish longcloth, western kersey, penistones, Devon dozens, plains.

The colonists had options for clothing themselves. Imports of British woolens were intermittently available. The latest fashions and the finest fabric came from Europe and those who could were happy to buy them. Oliver Dickerson writes in *Navigation Arts and the American Revolution,* "It just did not pay to produce cloth under American conditions when goods of as good or better quality could be had from abroad for less money." Labor was expensive in the colonies, and wool was often scarce.

Though many colonists relied on imports of both British woolens and linens, farmers were often hard pressed to buy imported cloth. Wool and flax both were widely grown and processed in the colonies. Even when the colonists made their own cloth, probate records showed that linen production outweighed wool by three to one. During trade interruptions home production was even more necessary.

At the same time that European settlers were importing wool fabric, Native Americans were also acquiring British woolen cloth and blankets, through trade for beaver and other skins. They appreciated wool for the same reasons people anywhere in the world did: it was light, it was warm when wet, it dried faster than animal skins, and it came in a pleasing variety of colors. Daniel Gookin, author of several books about "American Indians," described trade with Massachusetts Indians in 1674: "But, for the most part they sell the skins and furs to the English, Dutch, and French, and buy of them for clothing a kind of cloth called Duffils, or trucking cloth, about a yard and a half wide." From artifacts excavated in burial sites, trade with Native Americans can be dated from at least 1650.

By 1700, the population of the American colonies was 250,000, with more people living in New England than in the rest of the colonies. This balance would shift dramatically as the population of the colonies doubled each generation, until the first national census in 1790 revealed a population of 3,900,000. The textile needs of the growing population were great. The push and pull began between the motherland, interested in fostering the new market it had created for its woolens, and the colonials, the majority of whom were farmers with the knowledge, means, and incentive to make as much of their ordinary clothing as they could.

Domestic manufacture of wool was encouraged by colonial authorities. Beginning in 1645, the General Court of Massachusetts Bay passed the following order:

> Forasmuch as wollon cloth is so useful a commodity without which wee cannot so comfortably subsist in these parts, by reason of the cold winters, it being also, at present, very scarce and deare among us, and likely shortly to be so in parts from whence we can expect it,

> by reason of ye wars in Europe destroying in a great measure the flocks of sheepe amongst them . . . and wheras, through the want of woolen cloaths and stuffs, many pore people have suffered much could and hardship, ton ye impairing of their healths, and ye hazarding some of their lives . . . having an eye to the good of posterity, knowing how useful and necessary wollen cloths and stufs would be for our more comfortable cloathing . . . doth hereby desire all ye towns in general . . . that you will carefully indeavor the preservation and increase of such sheepe as they have already, and also to procure more with all convenient speed into their several towns by all such lawful wayes and means as God shall put into their hands.

This same act forbade the export of ewes or killing them before they were two years old.

Another order in 1656 required families to spin 3 lbs. fiber (linen or wool) 30 weeks out of the year, and an act in Massachusetts in 1664 required youths to learn to spin and weave.

Britain was distinctly uninterested in colonial manufacture, as it profited greatly from imports. In 1660 the British government attempted to protect the British woolen trade by forbidding export of wool and sheep from England. Later, the Woolens Act of 1699 would declare "wool or woolen goods could not be carried by ship or horse from one colony to another." Sheep were thriving on the islands off the coast of America, and the Act hoped to prevent the moving of sheep or woolen products to Long Island and Connecticut.

England's exports to the colonies increased over 50 percent between 1720 and 1770. Wool cloth was the major colonial import, keeping pace with the quadrupling of the American population. However, the volume of colonial manufactures also grew. This attracted the "suspicious Watchfulness" of agents of the British Government. Lord Cornbury, Governor of the Province of New York, wrote in a 1705 report to the British Board of Trade:

> Besides, the want of wherewithal to make returns to England sets men's wits to work, and that has put upon them a trade which, I am

sure, will hurt England in a little time; for I am well informed that upon Long Island and Connecticut they are setting upon a woolen manufacture, and I have seen serge, made upon Long Island that any man may wear. Now, if they begin to make serge, they will in time make coarse cloth, and then fine. How far this will be for the service of England, I submit to better judgements; but, however, I hope I may be pardoned if I declare my opinion to be that all these colloneys, which are but twigs belonging to the main tree—England—ought to be kept entirely dependent on and subservient to England, and that can never be if they are suffered to go on in the notions they have that, as they are Englishmen, so they may set up their same manufactures here as people may do in England . . .

In 1708 Caleb Heathcote wrote to the Board of Trade, "They were already so far advanced that three-fourths of the linen and woolen used, was made amongst them, especially the coarse sort; and if some speedy and effectual ways are not found to put a stop to it, they will carry it on a great deal further, and perhaps, in time very much to the prejudice of our manufactures at home." Also in 1708 a British surveyor, Bridger, wrote that "not one in 40 but wears his own carding and spinning."

Southern plantation owners, who by 1770 needed hundreds of yards of cloth each year to outfit their more than half a million slaves, ordered directly from England. Many plantations also had their own facilities for producing cloth. George Washington, who kept a weaving workshop at Mount Vernon, kept detailed accounts of the expenses incurred in weaving his own cloth between 1767 and 1771, with the intention of comparing home manufacture to imported goods. In 1768, he concluded it was not cost effective to make it himself, especially with falling tobacco prices affecting the balance of trade.

In 1765 with emotions running high over the Stamp Tax, colonists attempted a series of boycotts which evolved into a nonimportation movement, and translated into the wearing of homespun. It was a source of nationalist pride to wear homespun—cloth made in the colonies—rather than British cloth. The *Hampshire Gazette* from Amherst, Massachusetts, published an "Address to the Ladies" in 1769:

> *First then throw aside your high-top knots of pride*
> *Wear none but your own country linen;*
> *Of Economy boast,*
> *Let your pride be the most,*
> *To show clothes of your own make and spinning.*

In 1768, Harvard's senior class appeared for their graduation in homespun cloth of Rhode Island wool, and in 1769 Yale followed suit (as it were). In a diary entry from 1775, a young girl, Abigail Foote of Colchester, Connecticut, recorded "I carded two pounds of whole wool and felt Nationly."

At the first Continental Congress in 1774, an effort to avoid importations of British woolens was reflected in Article vii and Article viii.

> We will use our utmost endeavors to improve the breed of sheep and increase their number to the greatest extent; and to that end, we will kill them as seldom as need be, especially those of the most profitable kind; nor will we export any to the West Indies or elsewhere; and those of us who are, or may become, overstocked with, or can conveniently spare any sheep, will dispose of them to our neighbors, especially to the poorer sort, on moderate terms. (Article viii) We will, in our several stations, encourage frugality, economy and industry, and promote agriculture, arts, and the manufactures of this country, especially wool . . .

Certainly, going into the American Revolution, the amount of woolen cloth produced in the colonies was increasing. However, the population was also growing, and of necessity the colonists continued to import from Britain. Plains, a coarse woolen cloth, was imported from Wales and England for slave clothing. Needing 60,000 blankets for soldiers and lacking other alternatives, George Washington ordered them from abroad in 1775. Even bunting, a coarse wool fabric used in flags, was made in England. There was a cry to wear American homespun, but there simply wasn't enough of it yet.

Tools and Technology

The process of turning raw wool into a finished product involved a multitude of steps. It is safe to say that rarely did a single household do it all, debunking the myth of the self-sufficient colonial farmstead. Some households raised sheep and bartered their wool. Others had spinners whose yarn was either likewise bartered or sent out to be woven. In many cases the cloth would be sent to a fulling mill. Diaries, household accounts, account books of professional weavers, and merchant records show that any or all of the tasks—the work of preparing fiber and spinning, weaving, and especially fulling—were performed by different people, depending on family circumstances.

The Shearer

Illustration by Hilda Wilcox Phelps from *Texas Sheepman: The Reminiscences of Robert Maudslay,* edited by Winifred Kupper, University of Texas Press, 1951.

SHEARING, SKIRTING, SORTING

Sheep are typically shorn in the spring. After shearing, the fleece has to be skirted first, removing the shorter belly wool and dirtiest parts, and then sorted according to length and fineness. Wool coming from different parts of the sheep varies: the shoulders yield the longest and finest but belly wool is short.

Wool sorting, American Woolen Company, 1912.
LIBRARY OF CONGRESS

Sheep washing, from H. Stephens, *Book of the Farm*, 1889.

WASHING

Wool is usually cleaned, or scoured, before it can be carded and spun. A particular challenge in processing wool is that a newly shorn fleece is heavy with grease (natural oil that protects the wool) and suint (sheep sweat), constituting up to 50 to 80 percent of a fleece's weight. Both wash out with water; lanolin is the byproduct derived from the grease.

Sometimes the sheep were washed before shearing, usually in a stream or pool made by damming a stream. Washing the whole sheep was a British custom which served to remove surface dirt. The practice of washing sheep before shearing to remove at least surface dirt was difficult, and the necessity much debated. An observer noted, "The second reason is that it is wrong to require hired men to go into a brook and stand there all day for the purpose of washing the sheep. Now and then a man will protest against it and refuse to do it . . . ." The practice of sheep washing fell into disuse around the 1870s, when scouring machines, a succession of troughs through which the wool was soaked, then mechanically soaped, rinsed, moved, and squeezed dry, washed wool more effectively than when it was on the sheep's back.

A wet sheep had to dry before shearing, and also, although sheep washing might have served to remove mud, it did not wash the grease out sufficiently so that the wool would take dyes. A variety of agents were used in scouring. This description of one is from the *Domestic Manufacturer's Assistant,* one of the first tracts published for the woolen industry.

> In the first place fill the kettle two thirds full of water and one third of urine, that which is old if you can get it. You will then heat this liquour as warm as you can bear your hand in it for one or two seconds. Then put 5 or 6 lbs. of wool loosely into it . . .

Urine for this purpose was often collected in taverns and rooming-houses, with a preference being for the urine of beer drinkers. "That from persons living on plain diet is stronger and better than from luxurious livers. The cider and gin drinkers are considered to produce the worst, and the beer drinkers the best."

Wool can be sold washed or unwashed, but as wool loses so much of its weight when washed, buying grease (unwashed) wool required good guesswork since even minor miscalculation of shrinkage might make the difference between profitability or loss at a small mill. Scoured wool commanded more than twice the price of grease wool.

PICKING and CARDING

Picking and carding are both hand processes requiring more perseverance than skill, and were often performed by the very old and the very young. Even after washing, the wool likely contained vegetable matter and dirt that could be removed by pulling the locks of wool apart. Picking the wool before carding was an extra step, but would yield cleaner wool and make it easier to card. An article written later, in an 1852 issue of *Godey's Lady's Book,* told a story of times past, when a "picking bee" was held.

> Divested of their riding attire, they joined the circle in the general sitting-room, used for culinary, as well as other purposes, and there found the guests occupied in separating and cleansing tangled masses of wool . . . .

Hand cards for wool carding.
PHOTO BY JOANNE SEMANIE

Carding is the process that prepares wool for being spun into yarn. At this time, most yarn produced for use in the home was woolen yarn. Carding in the home was done with a pair of cards: wooden paddles with leather card cloth attached. The card cloth contains bent wires hand set into the leather. The wool is brushed between the cards to evenly distribute it. When finished, the wool is removed from the card so that it forms a roll of wool, called a rolag, or lamb's tail. When the spinner pulls from the end of the rolag, the fibers will be in an interlocking, airy mass.

The first carding machines were built in 1793, the second textile process to be mechanized after fulling. Mechanized carders were a series of cylinders which formed the wool into longer rolls of wool. Eventually these rolls would be spliced into continuous ropes, called a sliver.

SPINNING

Using a wool wheel (also called a walking wheel, great wheel, or high wheel) the spinner turns the wheel with the right hand while using the left hand to draft (pull) carded wool, with a twist being introduced off the end of the spindle (think of Sleeping Beauty pricking her finger up in the attic, but really, it's not that sharp), and simultaneously stretching her arm and walking backward away from the spindle as the wool takes the twist and becomes yarn. When an arm's length has been spun, the spinner then walks forward and winds the yarn up on the base of the spindle before starting over. Diaries show that whoever the spinner was in the family: farm daughter or matron, hired girl or slave, might typically spin 10 knots equaling one skein, or 800 yards, a day. The rule of thumb was that it took four spinners working full time to supply one weaver. However, most spinning was done around other tasks, perhaps only an hour out of the day. Various diarists recorded their daily output, some expressing satisfaction. It is hard to know now whether it was the satisfaction of accomplishment, or enjoyment of the task.

The first attempts to mechanize spinning were by Englishmen: Richard Arkwright, James Hargreaves, and Samuel Crompton. To understand the development of mechanized spinning, we have to look back at hand technology and the two types of hand spinning wheels:

WOOL
Colonial Period 1609–1780

Wool wheel, circa 1800. PHOTO BY JOANNE SEMANIE

Spinning jenny.

Spinning jack. From Baines, *History of the Cotton Manufactures in Great Britain*, 1835.

the wool wheel and the flax, or flyer spindle wheel. The wool wheel was a noncontinuous method of spinning, with the spinner drawing out and spinning 5 to 6 feet of yarn, at which point the spinner had to stop and wind it onto the spindle. The flyer spindle wheel simultaneously wound the spun yarn on to a bobbin as it was drawn out and twisted.

Arkwright developed a spinning frame, patented in 1769, based on the bobbin/flyer wheel. The Arkwright system consisted of a sliver of carded wool pulled between a series of rollers, simulating the motion of pulling between two fingers, after which it was twisted by the flyer and wound onto the bobbin. James Hargreaves patented the spinning jenny in 1770 which worked on the same principles as the wool wheel. Yarn was spun off a row of spindles, after which the spun yarn was wound onto the spindles, creating a cop. The invention of the spinning jenny would be followed by the spinning mule, invented by Samuel Crompton (no known relation to William Crompton, inventor of the first woolen loom, though both were from Lancashire) in 1779. The mule moved forward on a track to draw the sliver, and the yarn winding would be accomplished as the mule was pulled back. This was a muscle powered, hand driven machine until the application of belt driven power in 1795. The mule produced softer spun yarns, which were used for weft. The cops could be taken directly to be woven on looms.

PREPARING the WARP

Making warps was a skilled job requiring excellent math skills, usually performed by the weaver. To make a warp, the weaver had to wind the required number of warp ends, in whatever pattern had been chosen, to whatever the required length which could be anything from 50 to 100 yards. This was done by winding yarn on a warping board, until the required number of ends was wound.

LOOM SET UP

Loom set up required the weaver to first wind the warp on the back beam of the loom, and then draw in the warp ends through the harnesses and the reed.

Colonial Period 1609–1780

QUILL WINDING

Quill winding means winding bobbins for the weaving shuttle. Sometimes the wool wheel was used, sometimes a smaller version called a quill wheel was used which allowed the person winding to sit down.

Barn frame loom, circa 1800.
COLLECTION OF AUTHOR

WEAVING (Finally)

Weaving is most easily accomplished on a loom, a simple machine which raises alternate sets of threads (warp) to allow the passage of a shuttle carrying the weft. The typical handloom used for household production was a large timber frame, with four upright corner posts (sometimes referred to as the four posts of poverty) inside of which hung the working parts of the loom. The width of the loom was usually about 48 inches, a comfortable reach for a weaver to throw the shuttle from one side and catch it at the other, and suitable for weaving a cloth which might finish at 36 inches wide.

When the weaver presses the treadle a space opens between warp threads, called the shed. The weaver would then throw the shuttle into the shed, beat the weft thread with the overhanging beater, regulating the tension especially regarding the selvedges, and repeat the steps. When the weaving space in the front of the loom is filled, the weaver lets off unwoven warp from the back, or warp beam, and winds up the woven cloth on the front, or cloth beam. Weaving entailed many motions, beyond the simple throwing of the shuttle. For colored patterns such as checks or plaids, the weaver switches back and forth between multiple shuttles carrying different colors, exchanging them as the pattern dictates.

DYEING

Wool comes in different natural colors, from white to shades of gray and browns, all the way to deep chocolate. Wool accepts dye beautifully, as seen in the jewel-like tones in tapestries. It can be dyed at different points: dyed in the wool means it is dyed after scouring but before carding and spinning. This ensures consistent color through and through, and also allows for the mixing of colors during the carding process, one way to create a heather yarn. Wool can also be dyed after it is spun (skein dyed). Dyed yarns are used to weave stripes, plaids, and other patterns based on the warp and weft being different colors. Finally, wool can be piece dyed, meaning that yardage is dyed a solid color after it is woven into cloth. Early Americans used a variety of natural dyes to color wool including indigo, madder, walnut, and butternut. In the mid-nineteenth century synthetic dyes were developed and used in industry.

FULLING

Finishing woven cloth could be more or less involved, depending on the effect desired. After wool cloth comes off the loom, it needs to be washed to remove oil used in spinning and any residual lanolin. Beating or agitating it will accelerate the shrinkage into a tighter cloth. The wool fibers will interlace and fill out, and may be shrunk to various degrees by a combination of soap, agitation, and application of cold or hot water. Broadcloth requires fulling. At first, cloth was fulled by handsqueezing or foot stomping, often as a communal activity. The first fulling mill was established in Rowley, Massachusetts, in 1643, a town settled by 20 families from Yorkshire. Fulling mills were boxes containing hot water and soap, into which the cloth was folded. Water wheels were used to power hammers which pounded and rolled the cloth inside the box. From 1684 to 1689, 3,000 yards were processed annually. Many other fulling mills were built including one in North Brookfield, Massachusetts, which finished 5,000 yards of cloth in 1793.

FURTHER FINISHING
Brushing and Napping

Many fabrics, such as broadcloth, were brushed after being fulled to bring up the nap, and then sheared to create an even surface. This increased the insulation value of the fabric.

KNITTING

Knitting is another type of fabric construction, and requires very little equipment. The interlocking loops made with knitting needles make a stretchy fabric that was used in early days primarily for stockings. Stockings were available as imported items, but many households made their own. Knitting and spinning were often daily tasks performed by women and girls. Orphans were likely helpers to supplement a household's output, as a visitor observed about a Georgia orphans' house: "After dinner, they retired, the Boys to school and the Girls to their spinning and knitting." Slaves also were employed in these tasks.

Consumption of Domestically Produced and Imported Wool

Actual examples of cloth or clothing from this period are few. Articles of clothing surviving in museums tend to be atypical, exquisite heirlooms worn on special occasions, preserved and passed down in elite families. For example, there is a wool wedding dress of imported fabric in the Pocumtuck Valley Memorial Hall Museum in Deerfield, Massachusetts. Artifacts of ordinary clothing are scarcer. Homespun and handwoven fabric was so labor intensive to make that it was recycled and repurposed. It was common to recut clothes to fit a smaller person, or to take apart and turn a garment inside out so as to wear both sides, after which it might be destined for the rag bag and woven into a rag rug or sold to be recycled. Knitted stockings were a very common wool item, but even those might be taken apart and woven into rugs.

From diaries, autobiographies, account books, probate inventories, plantation records, ads for runaway slaves, and oral accounts as recorded in the Slave Narratives of the Federal Writers' Project, a laundry list can be compiled of types of domestically produced woolens and their probable uses. A list of wool cloth includes the simply named homespun, kersey, flannel, "Virginia cloth" (code for cloth used in slave clothing), and linsey-woolsey. These were all simple weaves, easily produced with the tools and technology widely owned by families at this time. Men's clothing made of these fabrics included coats, jackets, smocks, suits, trousers, and breeches while women's clothing included gowns, jackets, petticoats, and cloaks. Undergarments were likely made of wool flannel.

Woolens were not, of course, used solely for clothing. Most homes had wool blankets, two or three per bed. Some had wool curtains, as we know anecdotally from a diary that Samuel Sewell kept in which in describes a fire at his house and notes the "stubborn woolen window curtains

WOOL
Colonial Period 1609–1780

Homespun blankets, circa 1800.
COLLECTION OF AUTHOR

Linsey-woolsey bed curtains, circa 1750.
COPP FAMILY COLLECTION, COURTESY OF SMITHSONIAN INSTITUTION ARCHIVES

that refused to burn." Beds also could have curtains, many made of wool, which kept the bed warmer than the ambient temperature of the bedroom, which might be below freezing in cold weather. Bed curtains, for those who could afford them, entirely enclosed the bed, with a fabric roof, head cloth, and moveable sidepieces. Samuel Lane of Stratham, New Hampshire, gave each of his five daughters upon their marriages bed curtains, in addition to blankets and coverlids, as recorded in lists he made: "An Account of things I give to my daughter (name) toward her Portion."

Enslaved people's clothing was differentiated from freemen's, both in the articles of clothing worn and in the fabric used to make it. Men were issued jackets and trousers, women jackets and petticoats. Fabrics used in slave clothing were coarser, referred to as "slave cloth," "negro cloth,"

"Virginia cloth." Linsey-woolsey was used for work clothes. Written documentation of slave clothes included account books and slave narratives. Planter Robert Beverly kept an account book in which he records in 1768 that he bought for his slaves "100 yards plaiding and 60 ready made fearnaught waistcoats." Another planter's records show "four shirts, four pairs of pants (two coton and two wool)" as the annual issue for men.

It is difficult to quantify the amount of wool consumed during this period. One can come at it from the domestic production and foreign import angle—what might typical production have been in a household, as tallied in diaries or account books? Add to this merchants' trade records, showing how much imported fabric arrived. Or the consumption angle: looking at probate records and plantation records, working from how many items of clothing and household items such as blankets a person owned on average, then guessing how many yards of fabric went into each item? William Langdon, author of *Everyday Things in American Life, 1776–1876*, made a reasonable per capita guess of 3 pounds for the year 1776. Even for Gregory King, quantifying consumption would have been an impossible task, as impossible as Rumpelstiltskin's task of spinning a barnful of straw into gold. But it is certain that people depended on wool for survival in the New World.

NEW NATION
1780–1840

"In a country, the climate of which partakes of so considerable a proportion of winter, as that of a great part of the United States, the woolen branch cannot be regarded as inferior to any, which relates to the clothing of the inhabitants."

Alexander Hamilton, first secretary of the treasury,
Report on the Subject of Manufactures, 1791

Change was in the air, as the American Revolution ended with the signing of the Treaty of Paris in 1783, and the new nation endeavored to sort itself out and pay the debts it had incurred during the war. Generating revenue was critical. External revenue, import duties, were the source of 90 percent of government revenue in 1790, three-fourths of which was paid by Great Britain. In 1790 the federal government was also empowered to tax, after assuming the states' debts. Government and citizens looked for ways of increasing agricultural and manufacturing production in order to generate internal revenue.

In 1786 Tench Coxe, assistant secretary to the treasury under Alexander Hamilton, advocated for cotton cultivation, presciently anticipating that it might do for America what wool had done for England. He had seen it grown in Maryland, and concluded that it could be grown in the southern states. Up until then, the primary crop raised in the south had been tobacco, but falling tobacco prices may have prompted Coxe's meditations. Alternatively, taking a page from England's book, one patriot recommended wool growing as one way to pay the debt off, "setting Congress-men on wool-packs," asking money be put up to buy 6,000 sheep to be pastured around Philadelphia. The agricultural production of both wool and cotton grew substantially, but cotton would be the commodity that generated notable wealth.

The first national census in 1790 recorded a population of almost 4,000,000. Subsequent censuses showed the population roughly doubling every generation: from an estimated population of 2,700,000 in 1780 to 5,308,483 in 1800, doubling again to 9,638,453 in 1820, finally surging to 17,000,000 in 1840. With the Louisiana Purchase in 1803, the republic became, overnight, twice the size it had been. By 1818, the Flag Act presented a 20 star flag, as the country now extended west to include the new state of Illinois.

WOOL
New Nation 1780–1840

As during the colonial era, after independence American textile production was again inspired and influenced by the British textile industry. England, after centuries of handcraft woolen manufacture, was busy birthing the industrial manufacture of cotton, beginning with Arkwright's invention in 1769 of a spinning frame to make cotton yarn. The use of cotton, equally as much as the new technology, represented a revolution. It was a textile revolution, the popularity of cotton seemingly appearing from nowhere and soon overtaking woolen and linen cloth production. Inventions for spinning and weaving cotton were eventually applied to wool, so it is relevant to look at cotton technology.

Cultivation of cotton on a large scale in the United States began during this period. While cotton had been cultivated in Virginia since the days of John Smith, early on it was not as lucrative as tobacco. As late as 1791, its production was negligible. Experimentations with different varieties of cotton were begun in Georgia in the 1730s. But it was with Eli Whitney's invention of the cotton gin in 1794, that the mechanization of the cleaning of cotton made its production and processing affordable. Cotton production mushroomed from an unimpressive amount of cultivation in 1791, to 43.9 thousand tons sold to England in 1811, meeting more than half the needs in its cotton mills. Cotton cloth produced in England was exported back: 1802, exports of British made cotton had surpassed that of woolens.

Currier and Ives, 1880.
COURTESY OF AMERICAN ANTIQUARIAN SOCIETY

Satinet, late nineteenth century.
COURTESY OF THE JACKSON COUNTY HISTORICAL MUSEUM, MAQUOKETA, IOWA

During this period, the fabrics people wore were changing. To the flannels, kerseys, and linsey-woolseys of the colonial period were added new fabrics: linseys, satinet, and cotton Lowell cloth, as the United States made its first attempts at manufacturing. Clothing was becoming lighter: the frock coat appears in 1816 and trousers began to replace knee breeches around 1825. The availability of cotton yarn also led to the creation of new mixed fiber fabrics. New fashions were often made of satinet, a combination fabric made with a machine spun cotton warp and a wool weft. The example above shows the different surfaces of the fabric, where the wool weft shows on the front and the cotton warp on the back.

Populations, Human and Ovine

More waves of English immigrants swelled the population, as did Scotch-Irish, Scottish, Irish, and German immigrants, especially between 1820 and 1840. The 1790 census also included African American slaves, whose population grew from 575,420 in 1780 to 2,487,355 in 1840. Not included in the census until 1860 are Native Americans, but one estimate was 600,000 in 1800.

As the industrialization of cotton manufacture got underway in England, Britain passed stringent laws prohibiting people with knowledge of the new textile manufacturing processes to leave England; the laws also probihited the taking or sending of any models, drawing, or patterns. Fortunately for the American textile industry, a number of skilled mechanics made their way across the ocean. Most famous of these was Samuel Slater (emigrated in 1789), widely credited for setting up the first successful cotton spinning mill, in Pawtucket, Rhode Island; and the Yorkshire brothers Arthur and John Scholfield (emigrated 1793), inventors of carding machines.

James Beaumont, an accomplished kersey manufacturer from Yorkshire, knowing he was forbidden to leave England with textile equipment, packed casks of hardware in 1800 and masqueraded as a farmer's son going on a trading expedition. He set up a spinning mill in Canton, Massachusetts, in 1808 and wrote, "I then began to manufacture all wool cloth, careys, and satinets." He employed English handweavers to weave a wool fabric on a cotton warp, "beat them up well in the hand loom, so that when afterwards finished, you could scarcely tell the back side from the face."

Peter Dobson, a master spinner of Lancashire, arrived in 1809 and promptly set up a mill in Vernon, Connecticut. Wool was spun using a spinning jenny, and like Beaumont's factory, used handlooms for weaving. He and Beaumont were two of the first manufacturers of satinet in America, Mr. Dobson analyzing a sample of cloth and finding that "the warp was of cotton, five threads up and one down, and a filling of woolen yarn."

After the War of 1812 and a depression in England there was another wave of British immigrants between 1825 and 1835, some textile workers with experience in the new technology. Many of them left family in England, and we know of them and their subsequent adventures in the new world through their letters. At least ten members of the Hollingworth family of Huddersfield, Yorkshire: father George, children Edwin, Hannah, Jabez, James, John, and Joseph, and an aunt, uncle, and cousin arrived between 1826 and 1830, in the "company" pattern of emigrants 200 years before. The Hollingworths left an engaging account of their lives detailing work, moves, and family life. The most prolific writer was Joseph, the youngest son, who frequently "aided by his Poetical Muse," described happenings in verse. His versifying appeared to function as a way of letting off steam, as in a 1831 letter complaining about his employer:

> "For when men gets proud, and their power they abuse
> Then what resource have I, but to fly to my muse?"

Hollingsworth letter, July 17, 1831.
COURTESY OF THE AMERICAN TEXTILE HISTORY MUSEUM

From his letters we know that various family members were employed in Southbridge, Massachusetts, by Hamilton Woolen, then moved to Woodstock, Connecticut, where they leased and ran Muddy Brook Factory for three years. They made satinet on cotton looms, "stout iron-sided for cotton shirtings," that had been adapted for wool shuttles, showing how cotton technology was adapted for new fabrics. On Nov. 2, 1830, Joseph "finished 500 yards of satinet" and on Nov. 8, "finished 200 yds. more of satinet."

In another collection of letters, Edward Phillips wrote from Ohio in 1838, "I made a spinning machine, and looms, and buy wool and make it into cloth of different kinds." William Morris moved to Ohio in 1838, to Steubenville, where there were five woolen mills. Through these letters we can track what was being produced, and also observe the shift of population west in later decades. By 1820, 18 percent of the US population was living in the new states of Ohio, Kentucky, Tennessee, Indiana, and Illinois.

The population was moving west. One piece of evidence was how quickly the territories of Indiana and Illinois filled after they were admitted as states in 1816 and 1818 respectively. Increased population put pressure on to make it available for white settlement. At the end of 1829 President Andrew Jackson recommended removal of Native Americans, also referred to as consolidation. The Removal Act was passed in 1830, signaling formal intent of the government. Execution of the Act was haphazard and fragmentary. One example of the many treaties and arrangements was the Black Hawk Purchase, negotiated with the Sac and Fox Indians, which opened Iowa up to white settlement.

Ovine immigrants during the early 1800s were an essential element of the development of the American wool industry. In letters collected to inform Hamilton's *Report on Manufactures*, the state of domestic wool was summed up by a Mr. Jonathan Palmer of Stonington, Connecticut, who wrote to John Chester, "our Wollen Manufactury is wholly of the family kind and judge sufficient to furnish the Inhabitants, of there corse wairing aparel, those of a fine Quality are chiefly of foreign importation." Connor's report, "A Brief History of the Sheep Industry of the United States," states that farmers mostly kept sheep for domestic needs, with some wool left for sale or barter with local merchants. Average yield per sheep was 2 pounds of fleece.

After the haphazard importation of sheep during the colonial era, an effort to improve wool quality occurred when two fat tailed Tunisian Barbary rams were gifted to George Washington by the Bey of Tunis. The rams were placed with Judge Richard Peters of Belmont, Pennsylvania, who crossed them with Leicester and Southdown ewes. The result was Tunis, now an American heritage breed. Tunis are hardy, medium size sheep, and adapt to both northern cold and southern heat and humidity. They are raised for both meat and wool, with the fleece in the same fineness range as Corriedale and Shetland. Wool yield is on the low side by today's standards, only 4 to 5 pounds per ewe but an improvement over the 2 pounds of earlier breeds. Flocks were established in Pennsylvania, Maryland, Virginia, Georgia, North Carolina, and South Carolina. Tunis became the dominant breed in the mid-Atlantic and upper southern states until the Civil War, when they almost became extinct due to most of the stock being eaten by troops.

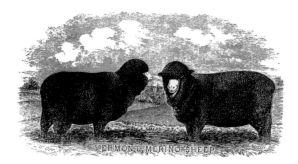

Merino sheep.

Other Americans interested in improving the quality of wool, and with connections in Europe, fixed their sights on the Merino, a breed known for its fine, dense wool. Additionally, Merino were good foragers, and appeared hardy as they migrated annually across Spain to the Pyrenees. In 1805 Du Pont de Nemours sent four rams to the DuPont family estate in Wilmington, Delaware, subsequently also the site of a woolen manufactury. Only one, Don Pedro survived the voyage. His washed fleece was reported to weigh 8½ pounds. DuPont is said to have had in 1812 "perhaps the largest and best" flock of sheep. Robert Livingston, Minister to France, acquired four sheep (two ewes, two rams) and sent them back to his home in Clermont, New York. In 1802 Colonel David Humphrey, Minister to Spain, sent a flock of 96 Merinos to his farm in Connecticut.[57] He also set up Humphreyville Manufacturing Co., Humphreyville, Connecticut, to card wool and later cloth.

However, until the Non-Importation Act of 1809, fine wool fabrics such as broadcloth were easily imported, so there was not much incentive to raise fine wool. The 1809 trade restrictions with Europe, and the ensuing war of 1812 finally spurred the home manufacture of fine wool fabrics. Robert Livingston, now back in the US and chancellor of the state of New York, had one of the largest flocks, 254 sheep, mostly not full-blooded Merino but demonstrating both better quantity and quality of wool. At the 1808 shearing, each sheep yielded between five and eight pounds of fleece. Livingston began his career of evangelizing for the Merino, publishing his "Essay on Sheep." Dispersing of the breed began. Elkanah Watson, a gentleman semiretired on an estate in Pittsfield, Massachusetts, and founder of the Berkshire Agricultural Society, purchased 40 rams from Livingston and placed at least one per township in the Berkshire area. At this time, rams could sell for $1,000-$1,500 each.

Spain had fiercely protected its fine wool sheep herds, but in the chaos of the Peninsular War between France and Spain beginning in 1807, and Napoleon's invasion, Spain chose to move its sheep out of harm's way. The American consul in Portugal, William Jarvis, and various American traders, bought almost 20,000 merino sheep and brought them to the United States in the years 1810 and 1811. Livingston acquired some of these. Following this major influx, Merinos spread from Maine to Georgia, west into the Ohio Valley, and to Indiana, Kentucky, and Illinois.

Upon returning to the United States in 1811, William Jarvis began raising Merinos in Vermont. Vermont eventually had the highest concentration of Merinos of all the states, peaking at 1.7 million in 1840. Thousands of acres of land were cleared in a very short amount of time to raise sheep in Vermont, New Hampshire, New York, and Massachusetts. Not long after, many of these farms were abandoned as soils were thin under the pastures and productivity declined. By 1850, there were only 1.5 million sheep in Vermont, New Hampshire, and Maine combined. The evidence of this land clearing would last for several hundred years, noticeable in the miles of stone walls which one still finds in the now reforested wilderness in these states.

Another feature of the Merino breed is the high content of yolk in the fleece. This yolk, or oil, washes out in scouring. Since wool was typically sold in the grease, a high grease content translates to more money when the wool is sold. Generally, wool loses half its weight when scoured, but Merino fleeces lost even more. Farmers in Vermont began to breed sheep with a lot of skin folds, which resulted in heavier fleeces (more surface area for the wool to grow on) and more grease. Ram's fleeces might weigh from 20 to 30 pounds when sheared, but only 7 to 8 pounds after washing. This engendered much acrimonious debate between wool sellers and buyers about regulating the accurate weight of wool, and the practice of breeding sheep with such excessive grease.

By 1812, there were ten million sheep in the US, and approximately 7 million people, or 1.4 sheep per person.

Politics and Wool

During the Revolutionary War, patriots attempted to wear domestically produced cloth (homespun) and boycotted British woolens whenever possible. The war ended in 1783; Washington was elected and inaugurated in 1789. At his inauguration, he wore a brown homespun suit of cloth made by the Hartford Manufactury. Manufacturing at this time was accomplished by handspinners working with carded wool roving to spin yarn to supply handweavers, and after weaving the cloth was finished in a fulling mill. In 1789 during this period of political regrouping, mindful of the new nation's goal to become self sufficient, John Bordley wrote "A Purport of a Letter on Sheep," opening with "The Increafing of wool, and to that end the enlarging flocks of fheep, ought to be a capital object with farmers in

America." He goes on to tell of his own experience of raising 130 ewes in Maryland, noting "they pafture through the Fummer, with little other attention to them than from now and then counting them." In his experience, he allows as how the estimates made of profits, can vary greatly and scarcely can two men be found to agree, depending on the attention and neglect, the manner of keeping them, and also differences in climate.

Once organized, the new House of Representatives asked the first Secretary of the Treasury, Alexander Hamilton, for "the means of promoting such as will tend to render the United States independent of foreign nations for military and other essential supplies." Gunpowder was on their minds, but so also was cloth for military uniforms. Hamilton set to work on his *Report on the Subject of Manufactures*, which was published in 1791. He observed that in 1781 the United States was not a manufacturing country, and set out to answer the question of whether manufacturers should be encouraged. In the first attempt to survey the resources and the industrial activities of the new nation, representatives of each state were asked to provide information on all areas of manufacture, from agriculture to iron working.

Arthur Cole, Harvard lecturer in business history and author of *American Wool Manufacture*, assembled the letters related to the collection of this information in *Industrial and Commercial Correspondence of Alexander Hamilton*, published in 1928. The correspondence is a treasure trove of information and opinions, if inclined to the anecdotal. But like other letters and diaries of the period, the correspondence provides a glimpse into the textile practices of the time. Here are a few excerpts from the letters:

Moses Brown wrote from Providence, Rhode Island, May 22, 1791, about his trials and tribulations in attempting to set up cotton spinning. This letter is particularly exciting, as without naming the "young man" we hear about the arrival on the scene of Samuel Slater, and the beginnings of industrial manufacturing of cotton.

> I attempted to set to Work by Water and made a Little yarn so as to Answer for warps, but being so Imperfect both as to the Quality and Quantity of the yarn that their progress was Suspended till I could

procure a person who had Wrought or seen them wrought in Europe for as yet we had not, late in the fall I recd a Letter from a young Man then lately Arived at Newyork from Arkwrights works in England informing me his scituation, that he could hear of no perpetual Spining Mills on the Continent but Mine & proposed to come and work them I wrote him & he came Accordingly, but on viewing the Mills he declined doing anything with them and proposed Making a New One using such parts of the Old as would answer.

In a letter dated Sept. 6, 1791, William Hillhouse of Montville, Connecticut, writes

I have not made Exact Calculations, but will submit for Once, that to Cloathe the People of this State only, it will require a number of Sheep to be shorn Yearly, not less than 100,000 [sic] What the number in the state was, the Last Deduction from the List I have forgot, but very short of the Number that is Wanted—You may I believe, often heard it mentioned that Two of the Most Material Objections to keeping Sheep are, In the first place Sheep are apt to be unruly and Troublesome, and in the Next place they will not pay the expence of keeping Equall to other stock, which is in a Measure True.

I know of no other remedy for the last objection but for the Good People in their Patroism to eat and Make away with as much Lamb and Mutton Sheep as possible instead of other Meat, which would make such Demand for Sheep as would induce the raising them, for it is a Well None fact, that the Wool that is Shorn from the Sheep is no Compensation to the Farmer for keeping them.

Another was sent Sept. 27, 1791, by Chauncey Whittelsey, Middleton, Connecticut:

I can only observe in general, That our farmers are mostly clothed by the produce of their farms, improved by the labour of their families; and these Woollens which are made in this (manner), are perhaps equal to any in the World, for the use of the day-Labourer. Considerable Quantities of Cloth are made for the purpose of barter; in that Way as well as for private Cloathing the Manufactures of this State, appear to be continually extending themselves.

A member of the Connecticut general assembly, Alexander King, wrote Sept. 12, 1791, from Suffield, Connecticut:

> The Wollen Manufacture is the Principal and the most Beneficial to the Inhabitants of any that is carried on in this Place—there are in Suffield about 400 families and about 5 Thousand grown Sheep, which will produce about 25 lb. To a Family on an Average this is all manufactured in the Domytic Way except Fulling and Dressing which is done at the Cloathiers Works.

From Pungoteague, Virginia, Robert Twiford wrote August 12, 1791:

> I have informed my self respecting the Manufactures as well as the length of time would allow, 45000 yd all yarn 30000d Cotton 45000 do wollen what we call Lintsewoose (linsey-woolsey) Flax linnen . . . I suppose that ¾ of the people are clothed in their own manufactury.

And finally a letter from Sept. 29, 1791, from Drury Ragsdale, who appears to have been more systematic in his surveying duties, undertaking to list all cloth manufactured in a year in King William County, Virginia, from the 20 families in his neighborhood:

> There being a scarcity of Wool it is generally mixed with Cotton, the warp being filled in with Wool that makes the cloathing of the Young and Domestic Negroes, and though not yeilding equal warmth with the cheap Kendal Cotton, is generally when wove double more durable than those Cottons.

After examining the accounts gathered, Hamilton offers the opinion that "the trade of a country which is both manufacturing and agricultural will be more lucrative and prosperous than that of a country which is merely agricultural." He further notes that "If there be anything in a remark often to be met with, namely, that there is, in the genius of the people of this country, a peculiar aptitude for mechanical improvements, it would operate as a forcible reason for giving opportunities to the exercise of that species of talent, by the propagation of manufactures."

Thinking even further ahead, Hamilton also writes, "The disturbed state of Europe inclining its citizens to emigration, the requisite workmen will be more easily acquired than at another time; and the effect of multiplying the opportunities of employment to those who emigrate, may be an increase of the number and extent of valuable additions to the population, arts, and industry, of the country."

Finally, in the section on wool, he recommends "measures, which should tend to promote an abundant supply of wool, of good quality, would probably afford the most efficacious aid that present circumstances permit. To encourage the raising and improving the breed of sheep, at home, would certainly be the most desirable expedient for that purpose; but it may not be alone sufficient, especially as it is, yet, a problem, whether our wool be capable of such a degree of improvement as to render it fit for the finer fabrics."

Not included in the survey was Native American manufacture. By this time, both the Pueblo and Navajo peoples had begun keeping sheep, and spinning and weaving wool. In *Pattie's Personal Narrative*, a journal kept by adventurer James Pattie, Pattie describes meeting a party of Navajo Indians in 1826, noting that "they have plenty of blankets of their own manufacturing . . . They dye the wool of different and bright colors, and stripe them with very neat Figures."

The Embargo Act in 1807 and the War of 1812 cut off English exports. With imported fabric unavailable, incentive for domestic manufacture was renewed, especially with the improvement in the wool supply thanks to the importation of Merino sheep.

At the end of the war in 1815 and the resumption of trade, there was a flood of imported cloth. Henry Clay championed the economic American system "to buy nothing from abroad which we cannot make at home, with due encouragement and protection from our government."

In an attempt to protect American manufacturers, the Tariff of 1816, the first duty levied on imported goods, was imposed by the government at rates of 20 to 25 percent. The tariff rate continued to rise in successive Tariff Acts: to 30 percent in 1824, 45 percent in 1828, and 50 percent in 1830. The tariff applied to both raw wool and finished cloth. In 1832, Louis McLane, secretary of the treasury, delivered the McLane Report, Documents Relative to the Manufactures in the US. He reported that since the tariffs of 1824 and 1828 had been enacted, American imports of British cloth had declined, but in the '30s they were increasing again. With so many different factors affecting the industry such as falling prices of finished cloth in England, rapid improvements in machinery, and raw wool prices in the US, the tariff was a crude instrument. The government would continue to adjust it.

Mechanical Innovation

In the colonial period, spinning and weaving were hand processes, using hand-operated equipment, resulting in a handcrafted product. Quality of the cloth depended on the skill of the spinner, weaver, fuller, and shearman. As processes began to be mechanized, the result was, at least in theory, a more standardized product.

In Britain, early attempts to mechanize textile processes sparked revolts. In America, without centuries of guild tradition and with scarcer labor, the mechanization of handcraft production was embraced, and aided by the relative ease of procuring patents for inventions. Patents, as presented in Article 1, section 8 of the Constitution, guaranteed "Authors and Inventors the exclusive Right to their respective Writings and Discoveries" and the Patent Office was formed in 1802. The textile field was wide open to innovation. American textile technology began with observations of British inventions, but soon leapfrogged into a genuinely American version of the industrial revolution.

Power loom weaving of woolens took place first in the United States, there being great opposition in England to mechanization of weaving woolens. Changes in textile technology were necessarily piecemeal, and each new innovation often came in response to an earlier one, building incrementally until all stages of taking fiber from its raw to finished state were mechanized.

Fulling, though the last step in making cloth, was the first process to be mechanized. Fulling was a fairly straightforward mechanical process accomplished by the application of agitation, soap, and water to the woven cloth.

Carding was the second textile process to be mechanized. Carding mills could take advantage of small drops on rivers, and thus often operating seasonally on streams which only ran high in the spring. Farmers and households bringing wool to carding mills to be carded had to first wash it themselves after shearing, but it was a time saver for families to use the mill and then take the wool rolls home to handspin, as machine spinning of wool had not yet been invented.

The Scholfield brothers, Arthur and John, came to the United States in 1793 from Yorkshire, England. By 1794, they were employed at the Newburyport Woolen Manufactury in Massachusetts. The first carding machine was 24 inches wide with a single cylinder, but the next model had two cylinders with a doffing comb. The doffing comb consisted of teeth set in a frame to remove the carded fibers from the cylinder. In 1798 the brothers moved to Montville, Connecticut, where they set up their own mill. In 1801, Arthur moved to Pittsfield, Massachusetts, opened a mill, and began building carding machines. An ad in the *Pittsfield Sun* ran the next year:

> ARTHUR SCHOLFIELD Respectfully informs the inhabitants of Pittsfield and the neighboring towns that he has a machine, half a mile west of the Meetinghouse, where he picks, greases, and cards wool into rolls on the following terms, viz; nine pence for one color and eleven pence for mixed, per pound. The wool is to be sorted and clipped, if necessary, and all the dung, burs, sticks, and other trash picked out, If that is not taken out, I will charge one penny per lb. more.

Scholfield Card.
COURTESY OF OLD STURBRIDGE VILLAGE

By 1804, Arthur was advertising carding machines for sale. These machines produced rolls of carded wool, relieving families from hand carding but not yet suitable to be spun by machine, as the rolls had to be pieced by hand.

In 1808, Arthur Scholfield manufactured 13 yards of broadcloth, and presented it to James Madison for his inaugural suit. The cloth was made from fine wool, and we can conjecture that it may have come from Elkanah Watson's purchase of Merinos from Livingston about the same time.

Carding mills using Arthur Scholfield's carding machines were set up in many Massachusetts towns: North Amherst, Hadley, North Hadley, Lenox, Williamstown, Cheshire, and Lee. By 1810 there were 1,776 carding machines in operation in New England, New York, and Pennsylvania. Carding mills needed water power, but could operate with even small drops on rivers, sometimes seasonally.

Spinning was the third textile process to be mechanized, with cotton yarn spun first in Samuel Slater's mill in 1790 using Arkwright-type machinery which was based on the flax wheel bobbin flyer principle of continuous spinning. In 1830, this machinery was redesigned into the forerunner of modern ring spinning, invented by John Thorp of Providence, Rhode Island. Danforth designed machinery in which the bobbin rotated instead of the spindle. Thorpe designed a ring traveller. Both improved the tensioning of the yarn. The early machines could only be used to spin cotton, and the product was a hard twisted yarn.

The spinning jenny and its successor in 1830, the spinning jack, operated on the principle of yarn being spun off the end of a spindle, as yarn was spun off the tip of the spindle on a great wheel. Both had multiple spindles and greatly improved productivity. The jack, based on Crompton's hand powered spinning mule, had a carriage that moved forward and back. Spinning on a spinning jack was a highly skilled job, as the spinner had to move the carriage to draft the yarn evenly. In 1825 a spinner using a jack with 140 spindles could spin 700 skeins per day, as compared to the 4 skeins of a handspinner.

Handspinning of wool continued for many decades throughout the evolution of the mechanization of spinning. Improvements continued to be made to spinning wheels in the home. Amos Miner, a wheelwright from Marcellus, New York, patented an accelerating wheel head in 1803 as a modification to wool wheels. The Miner's head was an intermediate pulley between the drive wheel and the spindle, placed such as to increase the speed of the spindle. It could be added to any existing wheel, with a few modifications. His timing couldn't have been better. Merino sheep had just been introduced to New York and neighboring states, carding mills were proliferating, and the Miner's head imparted more twist to yarn with less effort. When he first started manufacturing them, he made six to nine thousand a year. Tens of thousands of Miner's accelerating wheel heads would be sold in the decades that followed, with 62,000 being sold in 1865 alone.

In the early days of mechanization, only cotton could be spun strong and consistent enough to survive the harsh action of the early power looms. Weaving an all wool fabric awaited developments in wool carding and spinning. To fill the gap, American inventors first modified cotton looms to weave satinets, enlarging the shuttle boxes to accommodate the larger wool shuttles. The use of machine spun cotton warp made it less expensive, and stretched the supply of wool.

Satinet filled a niche for a moderately priced, serviceable cloth. Cole guesses that half of all woolen cloth produced in 1830 was satinet.

A Rhode Island mill owner, Zachariah Allen, went to England in 1825 and wrote back his surprise at finding no one there attempting to apply the new cotton weaving technology to wool as Americans had been doing for at least half a decade.

To weave cloth on a mechanized loom, five principal motions performed reflexively by handweavers needed to be integrated into the motion of the loom. These are shedding (lifting of the harnesses), picking (moving the shuttle across the warp, inserting weft), beating (the reed moving forward to push the weft into place), with concurrent motions of let off (a mechanism for unwinding the warp) and take up (winding cloth onto the cloth beam). All these separate motions needed to be timed to happen in an effective sequence; for example, the shed had to open fully before the shuttle could pass through it, and the shuttle had to pass all the way through before the beater could come forward. The accomplishment of these separate motions was the subject of much mechanical tinkering by early power loom mechanics. Early power looms looked much like hand looms, with leather straps to sling the shuttle from one side to another. The first power looms were complicated machines as they attempted to replicate all the actions of a human weaver, but in the end produced the plainest of cotton cloth. Later, woolen looms would benefit from all the innovations created for cotton looms, such as the Draper self-acting temple, patented in 1825. Temples look like a pair of alligator jaws through which the cloth rolls, on rows of spiked teeth angled outward. Temples keep the cloth stretched side to side as the beater comes forward to beat in the weft, and control the selvedges.

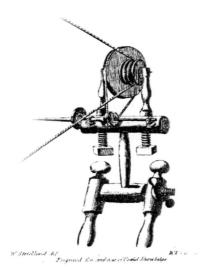

Miner's improved accelerating wheel head, 1811.
FROM *ARCHIVES OF USEFUL KNOWLEDGE* 2

In order to weave anything but relatively plain fabrics, a means of managing more harnesses had to be developed. In the 1820s, the dobby mechanism was invented. The dobby mechanism typically mounted on the side of a loom and transmitted design information from a pattern chain to the mechanism which lifted the harnesses. Early dobby chains were a series of wooden bars with metal pins. Each bar represented one throw of the shuttle, and caused harnesses to be lifted according to pins inserted in positions corresponding to each harness.

The Crompton Company manufactured the first loom that could accomplish all this. The founder of the company was an Englishman, William Crompton, who came to Massachusetts in 1836. By 1837 he was manufacturing cotton looms in Worcester, Massachusetts. In 1840, at the invitation of Mr. Samuel Lawrence, agent of the Middlesex Mills in Lowell, Crompton adapted his original loom to woolen weaving and wove fancy woolens at the Middlesex Mills. The original Crompton loom had one shuttle, 6 harnesses, and wove 40 inches wide at 45 picks per minute. It employed a metal pattern chain to lift the harnesses. Crompton's pattern chain design reigned as the industry standard for more than a hundred years, until it was replaced by computer assisted technology. It consists of metal bars built with risers and sinkers to move metal jacks which raise the harnesses. The new era of American woolen manufacture had begun. With William Crompton's 1837 loom, the American wool industry finally had the tools to compete with Britain's wool industry.

WOOL
New Nation 1780–1840

TOP to BOTTOM:
Patent model of Crompton's 1837 loom.
COURTESY OF ARCHIVES CENTER, NATIONAL MUSEUM OF AMERICAN HISTORY,
SMITHSONIAN INSTITUTION

Detail of harness lifting mechanism, 1837.
COURTESY OF ARCHIVES CENTER, NATIONAL MUSEUM OF AMERICAN HISTORY,
SMITHSONIAN INSTITUTION

Wool Cloth Manufacture

Until 1840, the manufacture of wool cloth continued on the parallel tracks of household and commercial manufacture. Household manufactures were stimulated by intermediate technological innovations, such as the Miner's accelerating wheel head and wool carding mills. Probate inventories from this period show still-significant numbers of spinning wheels, with regional variety. They were more common in rural areas.

Shaker communities, operating somewhere between household and factory scale, availed themselves of the new technologies for their cloth production. The Shakers of New Lebanon, New York, purchased a Scholfield card in 1809, and a "pleasant Spinner," presumably a spinning jenny, in 1824. The Shaker community of South Union Kentucky acquired a carding machine in 1819. Shakers sold products from their looms locally, as well as making cloth for their own use.

We can glean anecdotal knowledge of cloth production from diaries, merchant account books, probate inventories, and letters. For example, from a diary kept by Sarah Snell Bryant of Cummington, Massachusetts, we know that she had her wool carded at a mill in Chesterfield after 1800. Another diarist, Samantha Barrett, was a very prolific handweaver in New Hartford, Connecticut, and recorded in 1813 that she wove $57\frac{1}{2}$ yards of woolen cloth. The account book from the Shaker community in Canterbury, New Hampshire, shows they sold 60 yards of handwoven wool flannel in 1831.

Home manufacture was assisted by the publication of the first two pattern books written for weavers: Hargrove's *The Weavers Draft Book and Clothiers Assistant*, published in Maryland in 1792, followed by Bronson's *Domestic Manufacturer's Assistant*, published in 1817. Both are full of pattern drafts (threading and weaving instructions) for patterns common at the time such as dimity, satinet, birdseye, and herringbone. Bronson advertised "A plain system of directions" and gave detailed information on warp making as well as the drafts. Hargrove gives clear instruction on how to use the draft, for weavers who may have been unfamiliar with written drafts.

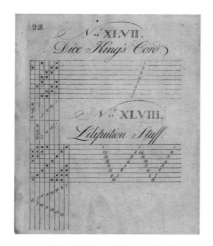

Page 22 from Hargrove's *Weavers Draft Book and Clothiers Assistant*, 1792.
COURTESY OF THE AMERICAN TEXTILE HISTORY MUSEUM

Hargrove's drafts are beautiful, resembling bars of music. In fact they were transcribed by an Alexander Ely, who also wrote musical notations for church songbooks.

Professional handweavers wove more specialized cloth, such as broadcloth which required a wider loom than those commonly found in homes. Commercial manufacture moved beyond the shops of professional weavers, to an early factory system in this period.

For example, in 1775 Pennsylvania patriots founded a society, "The United Company of Philadelphia for Promoting American Manufactures." At its inception, it featured a spinning machine, presumably a form of jenny with 24 spindles. However, it mostly relied on the handspinning of women, a "providential work, intended by the Almighty as a means of subsistence for such of his creatures among us as are not qualified to gain a livelihood any other way." Founded to manufacture woolens, cottons, and linens, operations were interrupted during the capture of Philadelphia by the British, but resumed in 1787 as the United Company of Philadelphia. Upon its revival, equipment included 4 spinning jennies. An article in the *New Hampshire Gazette* of 1788 states "cotton yarn, spun by the machines called jennies, is sold in Philadelphia at one dollar per pound."

The best known example of an early commercial enterprise was the Hartford Manufacturing Company, established in 1788. It attempted to make broadcloth to compete with the British import, a daunting proposition. Washington visited it and ordered cloth for his inaugural suit ("Hartford Gray"); Hamilton refers to the Hartford Manufacturing Company in his *Report on Manufactures* as a "precious embryo." As late as 1792, all the yarn was hand carded and handspun, although by 1795, also the year they ceased operation, their equipment included a locally made carding machine (preceding the Scholfield machine) and a spinning jenny. The width of the carding machine cylinder was only 16 inches. The period of manufacture also preceded the availability of domestic Merino wool, and so the wool used was imported from Spain.

By 1810, there were approximately 50 cotton spinning mills in New England. The mechanized spinning of cotton generated quantities of spun yarn, with weaving technology lagging behind. This resulted in an intermediate phase of textile production, and in new textiles. An outwork, or "putting out" system developed. "Putting out" was especially common in New England, where the informal transmission of textile craft had translated into a pool of women weavers, accustomed to making cloth for household needs. With an overabundance of machine spun cotton yarn, prewound warps were supplied to handweavers to weave cloth on handlooms, with the weavers paid by the piece. Putting out peaked around 1820, and may have employed as many as 12,000 weavers. This was a new source of income for women, in particular. Later, women were recruited to form the primary labor pool for Lowell's early mills.

Most woolen factories were relatively small affairs, and with the limited technology of the time and a shortage of experienced workers they prudently tended to specialize in a single type of fabric. Early factories sold their cloth locally and this simplification must have also worked for them as a marketing strategy. Others developed specialized markets such as the Hazard family's Peace Dale Manufacturing Company, makers of "negro cloth" in South Kingston, Rhode Island. The Peace Dale mill began with a carding machine in 1804, and by 1808 was manufacturing handspun and handwoven cloth with cotton warp and wool weft. In 1812 they acquired a power loom. In the 1820s Rowland Hazard travelled in the south, selling cloth and trying to ascertain needs. Beginning in 1822, Rowland Hazard's correspondence included invoices for bales of fabric destined for the slave trade, for example this letter dated "12 mo. 16th 1822"—

> Enclosed is an invoice for 6 bales linseys, 11 pieces 314 yds inferior kerseys, 346 yds plains, 321 yds kersey and kerseynettes

—and a letter dated "2nd mo. 20th 1825":

> The last year we have directed our attention to the manufacture of a very thick and strong kind of kersey milled and dyed of which about 15,000 yds were sold in Charlestown from 50 cts per yard and were approved by the planters.

The company acquired a sales agent, Mr. Ripley, based in Charleston, South Carolina. By 1828 they had power looms that could weave a yard wide. In 1855, the *Providence Journal* reported "the old mill, belonging to the Peace Dale Manufacturing Company, was entirely destroyed by fire early yesterday morning. The mill was making negro cloths and yarn for shawls." By this time, Rowland Hazard was active in the antislavery movement, securing the release of 100 men in New Orleans. A new mill was built, and the company appeared to have dropped the making of negro cloth, and moved into cassimere production.

What was being woven both in homes and in the new factories was a subject of great interest to states and the federal government. The first formal gathering of information on manufactures was undertaken in the census of 1810, more systemically collecting information than Hamilton's solicitation of state by state data for his 1790 *Report on Manufactures*. Some states were already doing their own surveys: for example the general assembly of Vermont in 1809 appointed a committee to prepare an account of manufactures of the state. That year, they recorded 1,043,540 yards of woolen goods produced, claiming it was almost sufficient for the household consumption. The 1840 census registered 96 woolen mills.

For the 1810 federal census, marshals asked the specific question, "How much fabric was made by families in the home?" Tench Coxe makes his appearance again, this time as executive bean counter, author of the Statement of the Arts and Manufactures of the United States, the first federal report of textile manufactures. Household manufacture of cloth was reported at 72,371,564 yards. From looking at the relative values of household and factory production, Coxe estimated that factories were producing less than 4 percent of the woolen goods made in the country in 1810. (In an aside, "Supplementary Observations," Coxe wrote, "No material, for cloths, for furniture and for apparel, is safe as wool in respect to fire . . . .") Coxe's figures agree with Rolla Tryon, author of *Household Manufacture,* whose assessment was that the years around 1810 were peak years for household textile manufacture. It is possible that the category of household manufacture may have included outwork weaving, which would have blurred the lines between household manufacture and factory work. Albert Gallatin, secretary of the treasury, would report to Congress in the Manufacturer's Report that almost all woolens were spun and woven in private families, though noting, "Water powered carding machines everywhere." The 1810 census also noted 325,392 looms, of which only 224 had fly shuttles.

Handspinning and handweaving persisted in inland and rural areas, and in the western states of Ohio, Kentucky, Tennessee, Indiana, and Illinois. An 1830 inventory in Stafford County, New Hampshire, showed that 82 percent of households inventoried listed spinning wheels, and 36 percent had looms. This would explain how an 1832 census in New Hampshire found that more than half of woolens were manufactured at home. New York State census surveys examined the household production statistics from 1825 to 1855, relative to population. Per capita allotment of all woolen cloth produced in New York households was 3.41 yards in 1820. Perhaps because of the greater availability of wool with the importation of Merinos, per capita wool consumption was up from previous decades.

The 4th national census of 1820 asked new questions: "Quantity and kind of machinery? Quantity in operation?" The American textile industry was still in its infancy: the census of 1820 revealed that only 1/3 of looms in cotton mills were power looms. One of the first large manufacturers, the Merrimac company, didn't build its first mill in Lowell, Massachusetts, until 1825, though by 1834 there were 19 mills.

The 1820 census reported 294 woolen firms, among them Stevens Mills in Andover, Massachusetts, whose founder began the manufacture of flannel in 1814. He brought women and handlooms into the mill for production. By 1827 the output of wool cloth in Massachusetts had risen to 9 to 10 million yards, of which a third was flannel and another third satinet. By 1830, "flannels now used are almost wholly American."

A census of the industry was taken in 1836 by Benton and Barry,

Statistical View of the Number of Sheep and an Account of the Principal Wool Manufactories. They found that four-fifths of the output of woolen mills was serviceable, simple to manufacture, medium-priced cloth, including cassimeres, flannels, satinets, and linseys.

Consumption

Southern laws required planters to provide clothing for their slaves; slaves comprised 35 to 40 percent of the southern population between 1800 and 1840. As slaves comprised a fifth of the US population during this period, and because clothing allotments were the main source of slave clothing, looking at slave cloth statistics is useful for thinking about wool consumption in general. While much was bought, either from manufacturers in New England or imported from England, quantities of cloth were produced on plantations. A conservative estimate of the woolen clothing needs of slaves was 5 yards per person (based on a common yearly allotment, to make one jacket, and either pants or petticoat). At this rate, the population of 2,784,000 slaves in 1840 would have required 13 million yards of fabric. Though there was some overlap with types of fabric worn by free men, fabrics for slave clothing included cotton such as Lowell cloth, combination fabrics (cotton warp/wool) such as plains and linseys, or osnaburg (cotton, tow linen) and woolens such as duffils and kerseys. Fabrics for slave clothing were manufactured specifically for slaves, baldly designated "negro cloth," and were generally designed to mark their status. "Negro cloth," such as the "Blue Niger" in the Noska draft book, most often referred to the coarser grade kerseys, and to plains and linseys, fabrics made primarily for slave outerwear. Wool used in the wefts was of a coarse quality, sometimes noted as being imported from Smyrna in what is present-day Turkey. Indeed, by laws enacted in various states in the south to regulate slave clothing, "negroes should be permitted to dress only in coarse stuffs such as coarse woolens or worsted stuffs for winter—Every distinction should be created between the whites and the negroes, calculated to make the latter feel the superiority of the former." As an example, Harriet Jacobs, a former slave wrote about 1826, "I have a vivid recollection of the linsey-woolsey dress given me every winter by Mrs. Flint. How I hated it! It was one of the badges of slavery."

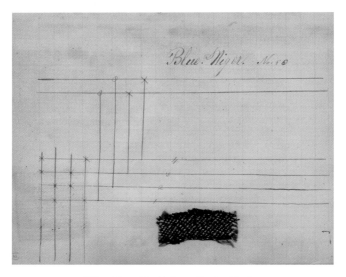

"Blue Niger Cloth" from Charles Noska
draft book, 1860–1867.
COURTESY OF THE AMERICAN TEXTILE HISTORY MUSEUM

The Federal Writers Project of 1936 to 1938 was a Works Progress Administration (WPA) initiative in which former slaves were asked a series of questions about life under slavery. Prompt #5 was, "What clothing did you wear in hot weather? Cold weather? On Sundays? Any shoes? Describe your wedding clothes." While the former slaves interviewed had all been born around 1850, later than the period we are currently considering, the clothing and method of manufacture was representative of earlier periods. Though the testimony varied depending on the situation each former slave had lived in, one can read into the replies that slave clothing was different from their owners', both in the articles of clothing and in the fabrics used.

A few samples:

> We ain't had nothing but the coarsest food and clothes.

> During hot weather we wore thin woolen clothes, the material being made on the farm from the wool of our sheep, in the winter we wore thicker clothes made on the farm by slaves. (Richard Macks)

Also included in the narratives were details about how the clothes were produced:

> Yes, we raised sheep—by the hundreds. The raw wool was first sent to the mill to be carded, then we would spin it.

> The sheep were sheared in May. The wool was washed, picked apart and combed and taken to the factory to be made into rolls.

> All of our clothing was homespun, our socks were knitted, and everything. We had our looms, and made our own suits, we also had wheels, and we carved, spun, and knitted. (Clayton Holbert)

> Each piece of (homemade) cloth contained 40 yards, and this cloth was used in making clothes for the servants. About half of the whole amount was this made at home; the remainder was bought, and as it was heavier it was used for winter clothing. (probably referring to imported woolen)

Again, it is impossible to definitively quantify wool consumption in this era, even as the Bureau of the Census began to collect data. There are a few troves of representative textiles, as in the inventory of the estate of Abraham Charles, who died in 1804, which included a "wollin shirt, a Great Coat, a Coat and wescoat (2), 2 pr of breeches, and an old coat." The Smithsonian holds a collection of textiles from the Copp family of Stonington, Connecticut. The collection includes clothing, sheets, coverlets, bed curtains, and tablecloths dating from 1750 to 1850. The

Copp family collection, 1890s.
COURTESY OF SMITHSONIAN INSTITUTION ARCHIVES

Copp collection is representative of textiles of the time, as it includes both homespun and imported materials. One of the family members, Daniel Copp, was a storekeeper whose advertisement in a 1798 newspaper also speaks to the variety of imported cloth available at the time. Working from a table of English cloth exports to the colonies, wool historian Elizabeth Hitz concluded that in 1790 per capita consumption of imported cloth was 3¼ to 5¼ square yards per adult male.

Perhaps the simplest solution for estimating per capita consumption is to think about the bare necessities. Both men and women minimally required five yards of wool cloth, making up either two pairs of wool pants and a coat, or a gown and a petticoat. Estimating consumption has to also factor in the very different widths that wool cloth was woven, whether it was narrow or broadcloth. For simplicity, let's use Cole's note[98] that wool averaged 13 ounces per square yard. Triangulating estimates by various historians and using several sources, including accounts of slave cloth, results in a convergence to somewhere around 4 pounds of wool per person in the late 1700s. In 1840, per capita consumption would rise to 4.5 pounds, perhaps reflecting the use of cotton warps in fabrics such as satinet.

INDUSTRIALIZATION and MECHANICAL INNOVATIONS
1840–1890

"O'er the forge's heat and ashes,
O'er the engine's iron head,
Where the rapid shuttle flashes,
And the spindle whirls its thread,
There is labor lowly tending
Each requirement of the hour:
There is genius still extending
Science and its world of power."

Charles Swain

By 1840, William Crompton's patent for a power loom signaled the beginning of industrial production of woolen cloth, and woolen mills joined cotton mills in the landscape. By then, cotton had achieved a textile revolution, the mills turning out millions of yards of white plain woven cloth, much of it to be later embellished after weaving by dyeing and printing. The weaving of woolen and worsted fabric was a smaller and more technical endeavor, with the styling done in the weaving, even if the pattern was just a simple twill.

Two American inventors, Elias Howe and Isaac Singer, patented their designs for sewing machines in the mid 1850s. Soon after the invention of the sewing machine, the factory system for manufacturing ready-made men's clothing developed. The combination of the mechanization of cloth production and innovations in sewing technology greatly expanded the garment industry. Americans of nearly all incomes could afford a greater variety of garments, and more of them. The availability of cheap cotton especially reconfigured cloth consumption and by 1850 many of these new clothes were made of cotton. The realization that clothes could be made more cheaply may have factored into the development of less expensive woolen fabrics, such as satinet and flannel. Per capita consumption of all fibers went up during this period.

Coinciding with increased supply was increased demand, as the population continued its rapid growth. By 1880 eighteen more states were added, bringing the total to 44. Pioneers flooded west with their oxen and wagons, and hopeful miners rushed to California in 1849. Immigrants continued to arrive from Europe. All needed provisioning, especially pioneer supplies such as ready-made clothes and blankets.

In 1861, Lincoln was inaugurated, the cotton states declared themselves the Confederate States of America, and the Civil War began.

Industrialization and Mechanical Innovations 1840–1890

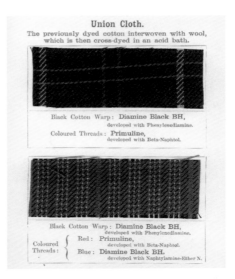

Union cloth (cotton and wool) samples, from 1895 dyebook *The Diamine Colors*.

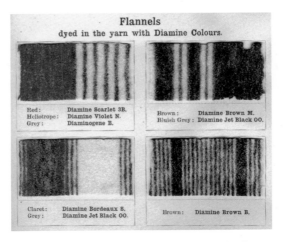

Wool flannel cloth samples, from 1895 dyebook *The Diamine Colors*.

An increased demand for wool uniforms and blankets gave the woolen industry an enormous boost, with many smaller cotton operations converting to wool to supply war needs.

Improved transportation helped nationalize the industry. Almost nine million pounds of wool was shipped east on the Erie Canal in 1850. In 1869 the transatlantic railroad was completed. Western wool could now be shipped to eastern factories, and both fabric and clothes were shipped west, disrupting local networks of cloth production in the Midwestern states.

Westward Ho

In the fifty years between 1840 and 1890, the population of the US almost quadrupled. New waves of immigration added another 15 million people (a quarter of the total): British, Irish, French Canadians, Germans, Scandinavians, Russians, and Italians. West of the Mississippi the new states of Iowa, Texas, California, Oregon, Nevada, Nebraska, Colorado, North Dakota, South Dakota, Montana, Washington, Idaho, and Wyoming were added. To empty this space for settlement, the American government embarked on a campaign to remove Native Americans. Non-native Americans and sheep moved into the new space, and the west emerged as the region where most of the wool was produced.

Congress authorized overland roads in the late 1830s, which helped to enable westward migration. One of the main routes west was the great Cumberland Road, running from Maryland to the Mississippi. Settlers were ferried across the Mississippi River to Iowa.

However, not everyone went west. Increases in the US population also resulted in the growth of urban areas. Urban population grew 4 percent every decade from 1840 to 1880. By 1880, the twelfth Census reported that half the population of the north Atlantic states lived in cities and large towns, increasing to 60 percent by 1890. In New England the rate was even higher. Farms were abandoned: 1,000 in Vermont, 1,300 in New Hampshire, 1,500 in Massachusetts, and 3,300 in Maine.

As new states and territories were opened up in the west, sheep moved with the settlers. Between 1840 and 1850, the west gained as fast as the east lost. The number of sheep doubled in Ohio in the 1840s and west of the Mississippi numbers of sheep increased rapidly. A report of the time likened it to a tornado. In 1844, 50,000 sheep were driven west into what would later become the prairie states of Iowa and Wisconsin. By 1860, the federal census reported 22,471,000 sheep, of which only a quarter were in New England and the Middle Atlantic states. Increased demand for wool during the Civil War led to an uptick in the number of sheep, up to 28,500,000 as reported by the 1870 census. Of these, 13 million were on midwestern farms. By 1870, Ohio, Michigan, Indiana, and Illinois were four of the nation's top six wool producers.

Huge flocks of sheep, as many as 25,000 in one instance, were driven to the gold fields to feed hungry miners, where they were sold for $12 to $15 each, a huge profit over the dollar a head price at home.

Churro sheep in Texas and New Mexico, offspring of the sheep brought by the Spanish around 1540, were used as foundation stock and bred with Merino. Merinos came to Texas with settlers of Austin in 1831. George Kendall, regarded as the father of the sheep business there, brought a flock of Merinos in 1856 to New Braunfels, Texas from New Hampshire. Many more sheep were driven to Texas in the 1850s; a newspaper, *The Valley Farmer*, reported in 1859 that "the state is not overrun with vicious and worthless dogs," a large impediment to sheep raising in the east. Eighty percent of the sheep in Texas were Merino or Merino grade. After the Civil War, in 1866, $2,000,000 worth of wool was shipped from Galveston.

Huge expanses were opened for grazing from Montana south to Texas by the 1862 Homestead Act. The Homestead Act gave settlers title to land after five years of residence and cultivation. One hundred sixty acres was the minimum allotment, not always enough land for grazing, particularly in arid areas. It was enough that an owner could claim a title for his ranch headquarters, though the sheep might be grazed off the ranch. The practice of sheep migration to seasonal pastures evolved. In California, summer grazing was in the Sierras. Sheep also found their way to western islands off the California coast: 50,000 on Santa Cruz, 15,000 on Santa Catalina, and 80,000 on Santa Rosa.

The completion of the transcontinental railway in 1869 opened the west to wool growers, as they could cheaply ship fleece east to the woolen mills. By 1885, about 45 percent of wool came from the far west. By 1890 50 percent of the sheep were in Montana, Wyoming, Idaho, Utah, Colorado, Arizona, New Mexico, and Nevada.

Illustration by Hilda Wilcox Phelps from *Texas Sheepman: The Reminiscences of Robert Maudslay*, edited by Winifred Kupper, University of Texas Press, 1951.

In 1882, Robert Maudslay was one of many Englishmen and Scots who came to Texas to try their hand at sheep raising, attracted by land companies' advertising campaign to bring settlers to west Texas. Maudslay worked variously as a sheepherder, sheep owner, rustler, and sheep shipper until 1905. He later wrote, "I had witnessed the closing years of the open range, and watched it year by year growing gradually smaller and smaller until at last it had vanished altogether."

Politics and Wool

The Civil War, like wars before and wars to come, stimulated woolen manufacturing. Wool was the fabric specified for uniforms, overcoats, and blankets, for its warmth, durability, and inflammability. In 1861, the Office of Army Clothing and Equipage in Philadelphia issued an invitation to mills to bid for army supplies. At the top of the list was wool: 50,000 blankets, and what looks like almost 20 yards per soldier of wool or

ARMY SUPPLIES—
OFFICE OF ARMY CLOTHING AND EQUIPAGE,
Philadelphia, May, 20th, 1861.
SEALED PROPOSALS are invited and will be received at this office, until 12 o'clock, M., on Monday, the third day of June, next, for furnishing by contract the following Army Supplies and Materials, deliverable at the United States Clothing and Equipage Depot, (Schuylkill Arsenal,) in quantities as required, viz:—

10,000 yards Cloth, dark blue, (indigo wool-dyed,) for caps, 54 inches wide, to weigh about 14 ounces per yard.
100,000 yards Cloth, dark blue, (indigo wool-dyed,) twilled, 54 inches wide, to weigh 21 ounces per yard.
130,000 yards Kersey, dark blue, (indigo wool-dyed) twilled, 54 inches wide, to weigh 22 ounces per yard.
175,000 yards Kersey, sky blue, (indigo wool-dyed) 54 inches wide, to weigh 22 ounces per yard.
50,000 army Blankets, wool, gray, (with the letters U. S. in black, 4 inches long, in the centre,) to be 7 feet long, and 5 feet 6 inches wide, to weigh 5 pounds each.
200,000 yards Flannel, dark blue, (indigo wool-dyed,) 54 inches wide, to weigh 10 ounces per yard.
100,000 yards Flannel, cotton and wool, dark blue, (indigo dyed,) to weigh 6½ ounces per yard.
400,000 yards Flannel, white, (cotton and wool,) 31 inches wide, to weigh 6½ ounces per yard.
400,000 yards Canton Flannel, 27 inches wide, to weigh 7 ounces per yard.
300,000 yards Cotton Drilling, unbleached, 27 inches wide, to weigh 6½ ounces per yard.
100,000 yards Cotton Drilling, unbleached, 36 inches wide, to weigh 8 ounces per yard.
200,000 pairs of half Stockings, gray, 3 sizes, properly made of good fleece wool, with double and twisted yarn to weigh 3 pounds per dozen pairs.
50,000 yards Russia Sheeting, 42 inches wide, best quality.
10,000 yards Brown Holland, 36 inches wide, best quality.
50,000 yards Cotton Muslin, unbleached, 36 inches wide.
20,000 yards Black Silesia, best quality, 36 inches wide.
4,000 yards Buckram, best quality, 40 inches wide.
8,000 sheets Wadding, cotton.
30,000 pieces Tape, (5 yards) white, ⅜ and ½ inches wide.
——Silk—red, white, yellow, green and blue, for flags, per yard.
——Silk twist and Sewing Silk, best quality, per pound.
5,000 Linen thread, W. B., No. 35 and 40, per pound.
8,000 Do.   do.   blue, No. 30 35 and 40,   do.

Advertisement for bids, 1861. Office of Army Clothing and Equipage. Army Supplies. Philadelphia, Pennsylvania, May 20, 1861, RHi X.
COURTESY THE RHODE ISLAND HISTORICAL SOCIETY

combination wool/cotton cloth for uniforms, underwear, and caps. The industry doubled between 1859 and 1869, many cotton mills converting to woolen production. One of the largest producers of woolens, the Pacific Mills of Lawrence, Massachusetts, grew from 1,000 looms in 1853 to 3,500 by the end of the Civil War. In 1864, 60 million pounds of woolens were produced for the army. With the invention of the sewing machine to help speed things along, "lucrative employment was given to a large number of hands, mostly American women."

The Civil War ended in 1865, leaving the nation with a national debt of four billion dollars. At this time the wool industry was one of the nation's most important sources of internal revenue (only exceeded by distilled liquor, iron, and tobacco). The number of sheep had expanded to meet Civil War needs; however, after the war and the end of the boom market there was a rapid depreciation of the value of wool. In addition, American wool growers were now competing with inexpensive wool from Australia and New Zealand. The National Wool Growers Association was formed in 1865. Their first report in 1866 noted, "Wool is an absolute necessary of life. In the climate of the United States it has, for the purposes of clothing, no attainable substitute." Their mission was to provide information to the United States Tariff and Revenue Commission to lobby for readjusting the tariff system toward rates more advantageous to American growers. The result of their lobbying was the Wool and Woolens Act of 1867, which established a high tariff rate on imported wool, protecting domestic wool growers.

At the first meeting of the Wool Growers Association, growers also took the opportunity to settle some long-standing issues amongst breeders: whether to wash the sheep before shearing; and also what to do about the excessive grease in Merino fleeces, resulting from breeding for deep skin folds around the neck to maximize the grease weight. Wool at this time was sold both unwashed and washed. Part of that debate is recorded in a special US Revenue Commission report, "Wool and the Manufacturers of Wool," where one exasperated observer noted, "Well, out of those ram's fleeces (Merino) that weigh from twenty-five to thirty pounds, the most cleansed wool that has been got has been some seven or eight pounds."

While in Congress, William McKinley raised protective tariffs, resulting in the McKinley Tariff of 1890 which instituted a duty of 50 percent on imported goods. (The Wool Growers offered McKinley an American-made and grown wool suit for his inauguration in 1896.) It was in the years between the Wool and Woolens Act and the McKinley Tariff that the country came closest to self-sufficiency in wool needs.

Native Americans were the only group that lost population during the period from 1840 to 1890, decreasing by more than half by the end of the century. The Indian wars of the Reconstruction era were the final act of government efforts towards Indian removal, also referred to euphemistically as "consolidation." In 1868, the Navajo signed a treaty, which granted them a reservation on their home land in the territory of New Mexico. Indian agent Theo Dodd, who took charge of the Navajos at Bosque Redondo that same year, sent an estimate of goods needed for the Navajos, which included 5,000 yards of linseys and 100 pairs red blankets. The Indian Service (part of the Department of the Interior) advertised and bought American made woolens as annuity blankets for Native Americans, part of a fixed annual compensation for them leaving their lands; in 1871 the US Army ordered 1,500 blankets for Indians on reservations. Annuity goods from 1868 through 1879 also included commercially spun woolen yarn. From 1868 (with an estimated 5,000 pounds of red and 1,000 pounds assorted other colors) until 1875, 60,000 pounds of 3-ply Germantown yarn from Pennsylvania were supplied to Navajo weavers.

In two years, 1863 and 1864, Kit Carson decimated the Navajo herds, obliterating Churro bloodlines. Dodd estimated in 1866 that there were only 1,100 sheep left, and recommended a restoration of at least 12,000. Beginning in 1869 the Indian Service issued two sheep or goats per person as part of the treaty settlement. By 1871, 30,000 sheep had been distributed, the breeds being a somewhat random collection of Churro, part Merino, and Shropshire. By 1878, there were 700,000 sheep on the Navajo reservation. The animals, along with the annuity yarn, enabled Navajo weavers to return to the craft of weaving.

Mechanical Innovation

In the early days of the industry, woolen mills either constructed their own machinery or adapted cotton machinery to suit their needs, as with satinet looms. Later, independent machine manufacturers such as the Crompton & Knowles loom builders, Whitin Machine Works (spinning), Davis and Furber (carding, spinning, and warping equipment), and Singer (sewing machines) established themselves as specialized builders of machinery for the textile industry. Their technological innovations made possible large scale production.

William Crompton, inventor of the first power loom that could weave fancy fabrics (more than 4 harnesses), founded Crompton Loom Works in 1940 in Worcester, Massachusetts. His loom employed a pattern chain, rather than cams, to effect the lifting of the harnesses. At the time, Jacquard and cam looms existed, but were not practical for production of fancy woolen cloth. Crompton's pattern chain mechanism reigned as the industry standard for more than a hundred years, until it was replaced

Portrait of William Crompton, circa 1840. FROM *100 YEARS OF PATENTS*

by computer assisted technology. It is comprised of metal bars which are built with risers and sinkers to move metal jacks which raise the harnesses. By 1857, a broad loom weaving 54 inches with 24 harnesses was weaving 85 picks per minute. The Congressional Committee on Patents, declared in 1878, "Upon the Crompton loom, or looms, based on it, is woven every yard of fancy cloth in the world." In 1880, the company developed their "improved broad fancy loom." These looms could not only weave patterns into the cloth, but could weave as many as seven colors in the weft, with extra shuttle boxes to hold different colors. The Crompton loom was so productive that by 1893, 75 percent of fancy woolen and worsted weaving was done on Crompton looms.

Meanwhile, another New England weaving entrepreneur, Lucius Knowles, had designed a loom which operated somewhat differently. His innovation was to put the harness and shuttle box pattern chains on the same shaft, so that they moved in a synchronized motion. He took out a patent in 1863 for a different concept also relating to the harness motion, an open shed mechanism. The Crompton loom was a "closed shed" loom, which meant that with every pick, every harness moved in order for the shed to reopen for the new pick. This required more power and caused more strain and wear on the warp. The Knowles loom was an "open shed" loom, which meant that individual harnesses would stay up until needed to move, causing less wear on the warp. His rival Crompton argued that the resulting differing degrees of strain on the warp threads would cause unevenness in the fabric. Eventually experiments showed that the warp threads were sufficiently elastic that this was not an issue.

In the early years of the woolen industry, wool was spun using spinning jacks. Jack spinners, men who muscled a spinning jack back and forth smoothly, demonstrated not only strength but great skill to avoid spinning twitty (unevenly spun) yarn. Around 1870, self powered spinning mules replaced the man powered jacks, and the cost of spinning was cut in half. As woolen spinning became mechanized, yarns began to be standardized. *Run* became the system used to describe the size of woolen yarn, replacing knots and skeins as spinning terminology. Mills designated the size of single spun woolen yarns as 1 run, 2 run, 3 run, etc., a run being equivalent to 1600 yards per pound.

IMPROVED BROAD LOOM, WITH HORIZONTAL HARNESS MOTION, ABOUT 1875

Crompton loom, 1875.

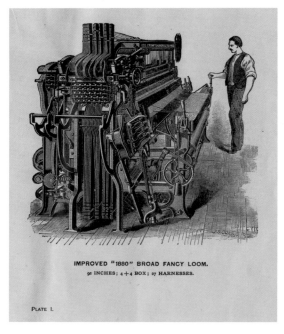

Crompton loom, 1880, trade catalog.
COURTESY OF THE AMERICAN TEXTILE HISTORY MUSEUM

Knowles loom, 1890s.

Knits as an industrially manufactured textile structure had been around since the patenting of the latch needle in Britain in 1849, which enabled the American inventor Isaac Lamb to build a flatbed knitting machine in 1863. Knitting machines were used to make articles of clothing that needed to fit closely, such as stockings and underwear. The Harvard Knitting Mill in Wakefield, Rhode Island, was one such knitting mill, making jersey undergarments and union suits in 1885.

Like developments in weaving technology, the invention of the sewing machine evolved incrementally, a combination of innovations and designs by numerous machinists. Elias Howe patented his 2 thread lockstitch machine in 1846, followed by Isaac Singer in 1851. Singer's machine featured a needle with pointed eye, a pressure foot to hold the fabric, and a foot treadle to power it. Allen Wilson patented a cloth feeding device in 1854. So many patents were held by different parties that a proposal was made in 1856 that the key players pool their patent rights, and license these to other companies wishing to use them. After the combination patent, the number of sewing machines manufactured increased dramatically. In 1880, Singer produced a half million machines in its Elizabethport factory in New Jersey.

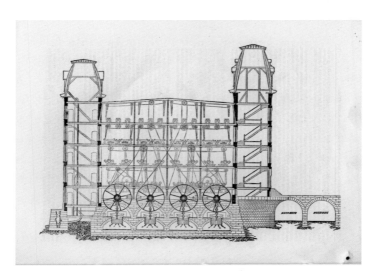

Manville Mill #3, from James Leffel & Co. trade catalog, 1875.
COURTESY OF THE AMERICAN TEXTILE HISTORY MUSEUM

It was not just the machines that enabled the Industrial Revolution, but the application of power to the machines. Textile machines—carders, spinning equipment, and looms—were connected to a central source of power, water at first, later steam, and finally electricity. A number of looms could be connected to the source of power by a wheel on the side of each loom. The wheel ran on a belt connected to an overhead line shaft. The overhead shafts were in turn connected to a power shaft which ran off the water wheel or turbine. While not strictly a textile innovation, when Paul Moody's belt drive system was introduced in 1828, replacing the British gear drive for transmitting motion from the main shafting of a mill, it greatly increased the speed of operation of textile machinery.

Woolen Industry

The first woolen mills were set up in New England, followed by mills in mid-Atlantic states. The Census of 1870 showed woolen mills in every state but Nebraska and Nevada, for a total of 2,000 woolen establishments with concentrations in New England (33 percent) and mid-Atlantic (27 percent).

Before railroads many woolen industries were started in the Midwest, keeping pace with settlers and meeting local demand. These pioneer manufacturers concentrated on cheap fabrics that sold well locally, primarily woolen apparel cloth for home manufacture of clothes such as flannel, cassimere, and woolen jeans for work clothing. Most marketed their product within a 50 mile radius, taking in fleece on a bartering basis. The mills were generally small operations, with second hand or improvised machinery making up 75 percent of the initial investment. One example was the Eagle Woolen Mill in Maquoketa, Iowa, started in 1874 by John Whitfield, an English immigrant. The Eagle Woolen Mill produced woolen fabrics, clothing, and blankets which were shipped throughout the region. They did commission weaving, including weaving for the Anamosa State Penitentiary in Iowa, until they closed in 1892.

The peak year in terms of the number of woolen mills was 1870: many smaller mills went out of business between 1870 and 1880, Midwestern mills in particular. Iowa is a good example: at the peak in 1870, there were 85 mills, by 1880, there were 34, a decrease of 51 or roughly 60 percent. Before railroads, woolen products from the east were more expensive than locally produced cloth. This price advantage protected Midwest mills. Eastern woolen manufacturers benefitted when railroad freight charges dropped during the 1870s and 1880s.

Also in the 1870s, worsted cloth became more popular with Americans. It was lighter in weight and dressier. Development of combing machinery facilitated the development of the worsted industry. Worsted wool spinning required entirely different equipment; instead of carding the wool had to be combed. It wasn't until the 1860s and 1870s that power driven combing machines were in place in American mills. Worsted cloth included serge for men and "stuff" for women. As worsted cloth became fashionable, people substituted it for woolens. In 1860, most of the long wool needed for worsted manufacture was imported from Canada.

Mills founded during this time came in all different sizes, and made a range of products. They included the Pacific Mill, one of the largest, in Lawrence, Massachusetts; the Woolrich Woolen Mill in Pennsylvania, founded in 1830 by John Rich, an English wool carder; the Troy Blanket Mill, started in Troy, New Hampshire, in 1865 to make horse blankets. Capps and Sons of Jacksonville, Illinois, who started in 1839 as a carding business, added looms in 1852 and made blankets and cloth during the Civil War. The Faribault Woolen Mill in Minnesota opened in 1865, known from the outset for its blankets. The Amana Society Woolen Mill, in Iowa, set up to make stocking yarn.

Hand Production

Even though commercially produced cloth was available, there is plenty of evidence that handspinning and handweaving persisted in rural and frontier areas through 1860. Pioneer woolen mills offered washing and carding services. The federal census of 1850 reports 31 carders in southeast Iowa alone, dropping to 25 in 1860. Handspun wool was used for making socks and other knitted items.

Women in both the North and South produced hand knit socks by the tens of thousands for soldiers during the Civil War. The US Sanitary Commission, precursor of the Red Cross, collected and distributed donated socks.

Handweaving of cloth decreased, but weaving of rag rugs at this time can be seen as evidence of a desire to handweave while producing a cheap saleable product.

Consumption

Wool and cotton coexisted in friendly competition during this period. Cotton far outsold wool; it was less expensive, easier to wash, and in addition to its use in apparel had wide household use for staples such as sheets and towels. Wool was used for more specialized purposes, and people valued it for its different properties. The expression "all wool and a yard wide" was coined during this period, appearing in print in 1866. It came to be used as a reference to a person's character, but generally was used to mean something of good quality. Taking the expression literally, we can infer that wool had a high value, and that wool as it came from a mill was literally expected to be 36 inches wide.

Per capita consumption of wool in the 1850s was 4 pounds, up from 3 pounds in the colonial period. Wool clothing included trousers, frock coats, and overcoats for men, and gowns, cloaks, and shawls for women, with flannel underwear and knitted stockings for both.

The transition of clothing construction from bespoke or tailor-made to ready-to-wear clothing originated with loose fitting clothes made up for sailors, miners, slaves, and soldiers. Ready-to-wear production began with the United States Army Clothing Establishment of Philadelphia, Pennsylvania, needing to produce uniforms for 10,000 men in the Union Army. One of the challenges of making ready-to-wear clothing is being able to make clothes that fit, with standardization of sizing. In order to achieve well fitting clothes, data was needed from actual humans in all their varied

Fall and Winter catalog, 1893–1894, Julius Saul. Warshaw Collection of Business Americana, COURTESY OF ARCHIVES CENTER, NATIONAL MUSEUM OF AMERICAN HISTORY, SMITHSONIAN INSTITUTION

sizes and dimensions. Information on men's body sizes was collected during the Civil War, when a million men were measured. Ready-made clothes spread into civilian wear: coats, pants, shirts for men, and cloaks for women. Wool flannel was widely used for undergarments, lining for overcoats, uniforms for soldiers and police, and coats for summer wear. By the 1860s, as mills began to make cassimeres, a finer cloth, satinet production declined.

Merchants found new ways of advertising their goods: in periodicals such as *Godey's Lady's Book*, *The Delineator*, and *Dorcas*, which featured patterns for home sewing and knitting. Department stores opened as urban populations grew: Marshall Field in Chicago (1852); Wanamaker's in Philadelphia (1862); Filene's in Boston (1881); and Lord & Taylor (1853), B. Altman (1865), Macy's (1858), and Bloomingdale's (1872) in New York. To serve rural populations, men's ready-to-wear clothes and wool fabric were sold in catalogs such as Montgomery Ward, beginning in 1872. Housewives could buy cloth from Montgomery Ward and sew clothes using Butterick sewing patterns (as seen in the *Delineator*).

Indianapolis, late 1890s. Warshaw Collection of Business Americana, COURTESY OF ARCHIVES CENTER, NATIONAL MUSEUM OF AMERICAN HISTORY, SMITHSONIAN INSTITUTION

LEFT:

The Delineator, Dec. 1892. Warshaw Collection of Business Americana, COURTESY OF ARCHIVES CENTER, NATIONAL MUSEUM OF AMERICAN HISTORY, SMITHSONIAN INSTITUTION

Children's clothes, 1882. Warshaw Collection of Business Americana, COURTESY OF ARCHIVES CENTER, NATIONAL MUSEUM OF AMERICAN HISTORY, SMITHSONIAN INSTITUTION

The years 1840 through 1890 encompass both slavery and the years after emancipation. Slave populations continued to increase between 1840 and 1860, continuing the purchase of cloth by slave owners. Post emancipation, participants in the Narratives, the WPA project that collected accounts from former slaves, mostly noted the lack of adequate clothing and the occasional trunks of donations for the freedmen sent from the north.

Henry and Nellie Warfield, Buckland,
Massachusetts, circa 1870s.
Photo by Jonas Patch

By 1880, per capita consumption of wool leaped to 8.5 pounds. This included imported wool, often as much as a quarter of what was used in manufacture, as there was never enough wool grown in the US to supply demand. Increased consumption may be accounted for by larger wardrobes, and the ease of acquiring clothing courtesy of the sewing machine, factory production, and ready made-clothing. In addition to clothing, wool consumption included bed blankets, horse blankets, carriage cloth, and felt hats. Textile mill production and the evolution of ready-made clothing production resulted in an abundance of woolen and worsted staples and luxuries that became part of life's necessities.

GOLDEN AGE of INDUSTRY
1890–1920

"An increase of nearly 20 per cent in population has been accompanied by an increase of nearly 30 per cent in wool consumption — a comparison that speaks for itself."

US Treasury Department, *Wool and Manufactures of Wool*, 1894

The period 1890 to 1920 was the high point in both production and consumption of wool in America. A hundred years ago wool could be found in all parts of the wardrobe, from underwear to overcoats. Women's fall and winter dresses were made of wool; every man owned a wool suit. Both men and women wore wool overcoats. Before central heating, the first line of defense against the winter was woolen long underwear. Wool fabrics specified in clothing from 1890 through 1920 included broadcloth, cheviot, flannel, gabardine, serge, tweed, and worsted. Per capita consumption was at an all-time high of 9 pounds in 1890, then dropping to 8 pounds in 1900, quickly sliding to 3.4 pounds in 1920, as lifestyle changes caught up with population.

Numerous interlinking forces drove the golden age of industry. First, fueling the demand for everything consumable was the exponential increase in the US population, doubling again between 1890 and 1930. It was a seller's market for consumer goods. Sales of consumer goods in general tripled between 1909 and 1929. Demographics were shifting, with the greatest population increase affecting urban areas. Second, ready-to-wear clothing was becoming widely available. The increase in quantity and decrease in cost of ready-to-wear clothing was made possible in large part by the immigrant labor arriving in New York. Third, sales of ready-made clothing were assisted by developments in retail merchandising: department stores, women's magazines, and advertising. The heyday of the department store coincided with urbanization of the population. Finally, the woolen and worsted industry benefitted from greater restrictions on imports in the form of tariffs. The Dingley Tariff of 1897 restored protection to woolen manufacturers and the decade of most rapid growth, 1900 through 1910, corresponds to the most restrictions on imports. As a result, the value of products of woolen and worsted mills rose from $2,200,000 in 1890 to $10,654,000 in 1920. Much of this increase can be attributed to the increase in worsted fabric manufacturing, a product that previously had been imported but which could now be made in the US due to advances in wool combing (preparation for spinning) technology. Worsted fabrics replaced woolen for dress goods and men's suiting. Developments in weaving technology increased production, and consolidation of companies such as Crompton & Knowles and the American Woolen Company increased their competitive advantage.

The sheep continued to trot west. In 1890, there were about 44 million sheep in the US, half of them in western states. Twenty years later two-thirds of the sheep would be west of Missouri: in Texas, the Rocky Mountain States, and the Pacific coast. New breeds were developed for western sheep farmers, who were interested in a sheep that could be used for both meat and wool. The Corriedale, an Australian crossbreed of Lincoln and Merino, made its appearance in the US in 1914. New American breeds included Columbia (1916), Targhee (1926), and Romeldale (1915).

The sheep population decreased to 39 million by 1900, and decreased again before World War I. This was partly due to removal of the tariff in 1913, making foreign (Australian) wool free of import duty. At the same time production of woolen goods was increasing, which meant that a good percentage of wool had to be imported. William Wood of the American Woolen company lent his voice to the effort to increase flocks, saying "though the population of the United States is rapidly increasing, the number of sheep and the amount of wool produced here are falling off."

World War I reminded everyone just how important wool was in wartime, and buoyed the wool industry as it had in earlier wars. Even before the US entered the war in 1917, the American Woolen Company received orders for uniforms and blankets from the Allied governments, where industrial production had ceased because of active fighting. American Woolen Company orders were up 40 percent in 1916 from 1915. In 1917, American Woolen landed the largest contract for textiles that the US government had ever awarded, $50,000,000.

During World War I, the White House pastured Shropshire sheep on the south lawn as a symbol of home front support of the troops overseas.

Sheep on the White House lawn, 1918.
LIBRARY OF CONGRESS

Feeding lambs, USDA publicity shot, 1916; the people are unidentified.
COURTESY OF NATIONAL ARCHIVES AND RECORDS CENTER

Shropshire, a British black-faced mutton sheep with medium grade wool was a popular breed at the time in the US. The flock, numbering 48 at its peak, saved manpower by cutting the grass and earned $52,823 for the Red Cross through an auction of their wool. The Red Cross also supported the war effort by launching a knitting marathon of hand knitted items to supplement the government supply: distributing yarn and instructions and eventually collecting almost 24 million hand knitted garments. Knitters took up their needles, though socks were also knit on the new hand cranked sock knitting machines. Gearhart, one of the companies manufacturing these machines, supplied one to all the Red Cross service centers.

Sock knitting machine, from Auto-Knitter Instruction booklet, 1924.

Demographics and Ready-to-Wear

The trend towards urbanization continued, as by 1910, one-fifth of the population lived in cities of 100,000 people. There were 200,000 African Americans who moved north in the Great Migration between 1890 and 1900; they were just one sector of the new urban population. From 1910 to 1920 another 300,000 to a million African Americans followed. Almost all went to industrial cities, specifically Detroit, Cleveland, Chicago, New York, Cincinnati, and Pittsburgh, where they found employment with factories, coal companies, railroads, and steel mills.

Increases in the US population resulted in part from another great wave of immigration, as the country absorbed 23 million immigrants between 1880 and 1920. The Immigrant Act of 1924 would be the first numerical restriction of immigrants, effectively ending mass immigration, but from 1880 to 1920, primarily eastern and southern Europeans came to the US. They filled jobs that industrialization had produced, especially in the textile and ready-made clothing industries. Between 1890 and 1920, manufacturing employment increased by 150 percent. Immigrants would fill half of those jobs, especially in the urban areas of the Northeast. Just as English immigrants in the early 1800s had supplied skills needed to furnish the mills and operate the factories in the early days of the American textile revolution, Italian and Russian immigrants familiar with clothing construction filled the lower East Side in New York and powered the garment industry. Manufacturers of clothing had their clothing assembled by contractors, one manufacturer sometimes working with hundreds of small shops. The contractors relied heavily on their social networks, relatives, and hometown people to fill needed positions. With a ready supply of labor, the sources for fabric, and the New England woolen and mid-Atlantic worsted mills, New York became a fashion center. The New York garment trade produced half of all ready-made clothing in 1900.

Mass production of clothing, with immigrant labor arriving to work in the ready-to-wear clothing industry, resulted in falling clothing prices, making clothing more affordable. In 1915, when 96 percent of annual family incomes were less than $2,000, a man could buy a wool suit for $15. Prices fell due to cheap labor as well as the advent of the factory and the assembly line for constructing garments. Garments were constructed in many separate operations, and these "deskilled" jobs could be learned quickly even by inexperienced operators. Speed in garment construction was also facilitated by the electrification of and technological advancements in fabric cutting, sewing, and pressing. Professor Jesse Eliphalet Pope wrote in the *Clothing Industry of New York*, 1905, "no single industry has done so much to administer to the happiness and well-being of the masses; for it has furnished, at a cost within the reach of all, an abundance of one of the prime necessities of life."

Ready-to-wear women's clothing lagged decades behind men's. The Sears & Roebuck catalogue offered nothing for women in 1894; but in 1910 they added women's suits in sizes 32 to 44, with a proportional choice of small, misses, or stout. By 1920 the catalogue's women's wear section took up 90 pages. Women's ready-to-wear was aided by a change of styling to "separates," skirts, and shirtwaists. However, enough data to create sizing standards for women was not definitively available until 1939–1940, when a Works Project Administration project undertaken by the National Bureau of Home Economics (a division of the US Department of Agriculture) measured 15,000 women.

The availability of immigrant labor, the application of the factory system to making clothes, the effects of factory systems on clothing prices, and the mass marketing of clothes all contributed to a greater consumption of clothes.

Cotton and Wool: The First Interfiber Competition

In 1900, the value of goods produced by the cotton industry was ten times that of the woolen and worsted industry. People wore cotton dresses, shirts, jeans, (patented by Levi Strauss in 1871), and summer underwear. Cotton was a general purpose fiber, more easily laundered. Doing the laundry was a monumental household task, with an entire day a week, traditionally Monday, being set aside in many households. As the germ theory of disease was propounded in the 1890s, sanitarians working to improve public health urged people to launder and change clothes regularly. Cornell University's home economists published a series of bulletins for the Farmers' Wives Reading Course. One bulletin, titled "The Laundry," recommended boiling underclothes for ten minutes. This was easier to do with cotton, as boiling causes wool to felt and constrict to a unusable fraction of its former self.

However, the relationship between the two fibers was a friendly one. Wool was purchased for different reasons than cotton, with specialized applications based on its versatility. Though the pioneer era had passed, wool still appeared everywhere in men's, women's, and children's wardrobes. Wool flannel, a lightweight woven fabric was the standard material from which winter camisoles, undershirts, petticoats and under-

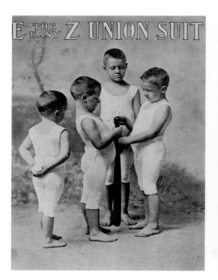 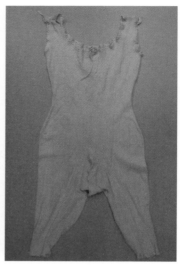

LEFT:
Union Suit, from postcard, 1915.
PHOTO BY WILLS WHITE, COURTESY OF *IMAGES FROM THE PAST*, BENNINGTON, VT.

RIGHT:
Woman's onesie, Dr. Jaeger's Sanitary Woolen System, 1911 catalog.
COURTESY OF THE AMERICAN TEXTILE HISTORY MUSEUM

shorts were fashioned. By the late 1800s, the technology had evolved to make fine knitted wool fabric as well as stockings. In 1883 the Jaeger Company developed "Dr. Jaegers Sanatory Woolen System," a line of porridge-colored knitted woolen underwear, with sales literature that read, "The evil effects upon health, due to the material and form of the ordinary clothing, were hardly suspected until D. Gustav Jaeger, of Stuttgart, began to practice his investigation on the subject." His investigations turned up information that "woolen clothing exerts its beneficial influences by virtue of its non-conductivity of heat and electricity, its permeability to moisture, its attraction for air, and its repulsion for water." In addition, Dr. Jaeger claimed that by wearing of close fitting woolen clothing he could conquer corpulence. Under "the Cleanliness of the Sanatory Woolen System" woolen clothes merely needed to be brushed, not washed because "the body collects less dirt in the SWS [Sanatory Woolen System] than in cotton or linen."

The Sanatory woolen shirt did what wool was known for: absorbing moisture and keeping a person warm. By 1911, the SWS was sold in stores in New York City, Brooklyn, Chicago, Boston, and Philadelphia.

Fascinating Widow movie, 1912. Warshaw Collection of Business Americana, COURTESY OF ARCHIVES CENTER, NATIONAL MUSEUM OF AMERICAN HISTORY, SMITHSONIAN INSTITUTION

In May 1917, the American Association of Park Superintendents published bathing suit regulations to ensure modesty. Among other things, the regulations specified that there be no white or flesh colored suits, and that men's suits required a skirt. Bathing suits which had begun as truncated union suits were restyled. Knitting mills, originally founded to make socks and wool jersey undergarments, reinvented themselves to fashion wool bathing suits. One such company was the Portland Knitting Company. It was founded in 1910 by Jon Zehntbauer and Carl

Jantzen. Carl Jantzen invented an automatic circular knitting machine, reducing the cost of knitted fabric. In 1913 a rower came in looking for trunks appropriate for rowing, which the mill designed using an elastic rib stitch. In 1918, the company changed its name to Jantzen, and set about becoming a pioneer in wool bathing suits.

LEFT:
Tennis costume, 1880s.
RIGHT:
Mail order wardrobe, 1890s. Warshaw Collection of Business Americana,
COURTESY OF ARCHIVES CENTER, NATIONAL MUSEUM OF AMERICAN HISTORY, SMITHSONIAN INSTITUTION

During these years, women's wool clothing included suits, cloaks, dresses (consisting of two pieces), tailored suits, and skirts. Skirts were long and under the skirts in the winter women wore petticoats and a union suit. Everyday wear was supplemented by special clothes for activities such as riding, tennis, and bicycling.

Serge sample, 1895. From 1895 dyebook *The Diamine Colors*.

In the men's department, ready-to-wear clothing and wool consumption converged in the ubiquitous blue serge suit made of worsted wool. Serge is a twilled fabric which because of its harder twisted worsted warp yarn, wears well and was often used for military uniforms. Men's winter pants were also wool. A new category of clothes was added to menswear: sportswear. The Jordan Marsh department store in New England sold golf and cycling suits in 1897, an example of the new clothes specially designed for sports.

For both men and women, the estimate of 9 pounds of wool per capita can be accounted for fairly easily when considering the amount of wool need to produce a new suit, two new dresses, a wool skirt or pair of pants, and long underwear. Weights of clothing then were much heavier than they are today. A lady's walking costume of the late 1800s required 9½ square yards of fabric, and skirts in 1890 had a 4½ to 5 yard circumference. Add in two wool items that are not so common today: blankets and overcoats. A long wool overcoat could weigh as much or even more than a blanket, and a wool bed blanket weighs 4 to 5 pounds, depending on the size. Of course, wool blankets and coats both last for more than a year, so an investment in one must be amortized over years.

Men's wool overcoats, Kirschbaum Company, 1912. Warshaw Collection of Business Americana, COURTESY OF ARCHIVES CENTER, NATIONAL MUSEUM OF AMERICAN HISTORY, SMITHSONIAN INSTITUTION

Man wearing suit, carrying overcoat, 1920. PHOTOGRAPH BY ADDISON SCURLOCK, COURTESY OF SMITHSONIAN ARCHIVES CENTER

WOOL
Golden Age of Industry 1890–1920

A number of mills specialized in blankets, which required wider looms. In 1890, the Elkin Valley Woolen Mill in Elkin, North Carolina, changed its name to the Chatham Manufacturing Company. Around 1910, they were the largest blanket manufacturer in the South, making millions of blankets during each of the World Wars. Pendleton Woolen Mills was founded in 1895 to make blankets for the Umatilla and Cayuse tribes in Pendleton, Oregon as annuity allotments. The blankets were initially sent to reservations, but the blankets soon became popular with tourists. The company would become famous for their "Indian pattern" camp blankets, Jacquard woven designs referencing Native American motifs.

Ellie Ganado. Photo by Fred Harvey, 1906.
LIBRARY OF CONGRESS

Tom Ganado. Photo by Fred Harvey, 1906.
LIBRARY OF CONGRESS

Department Stores and Other Delights

Up until 1876, when John Wanamaker converted a former railroad depot in Philadelphia into the first department store, clothes and other necessities were sold in dry goods stores, usually small employer-owned businesses. However, to move the mountain of goods made possible by mass production, something beyond the normal mechanism of supply and demand was necessary. Wanamaker helped in the creation of the "land of desire," a world in which consumers were enticed into wanting goods. Katherine Fisher, an early advertising expert, wrote in 1894, "Without wants, no demand to have them supplied." Sales of clothing were assisted by developments in marketing: department stores, mail-order catalogs for those not living in proximity to stores, and magazine advertising. From the 1890s on, department stores such as Wanamaker's (Philadelphia), Marshall Field's (Chicago), Macy's (New York), and Filene's (Boston) concentrated in urban areas and offered inducements such as installment buying, free delivery within city limits, and return policies that guaranteed "satisfaction or your money back." They sold appliances, housewares, and ready-made clothes. The grandest ones became palaces of consumption, with lavish displays and fashion shows. Serving rural customers, Montgomery Ward of Chicago put out their first catalog in 1872, and Sears and Roebuck, also of Chicago, followed in 1893. Sales in department stores doubled from 1914 to 1927.

Women's and men's magazines both featured fashion articles and clothing advertisements. A number of women's magazines were started between 1873 and 1910: *The Delineator* (1873), *Dorcas* (1880), *Good Housekeeping* (1885), the *Ladies' Home Journal* (1886), *Woman's Home Companion* (1897), and *McCall's* (1890). *The Delineator* was published by the Butterick Publishing Company, and featured their patterns prominently. Its publishers claimed, "*The Delineator* was the first magazine in the world to give expert information on the problems of food, sanitation, clothing, homebuilding and the scientific care of babies." Ready-to-wear clothes featured prominently in both advertising and feature articles. *Dorcas,* billing itself "Magazine of Women's Handiwork," was published monthly, with detailed knitting patterns as well as poetry and regular columns such as "Helps for Homely Women." Dorcas was a yarn manufacturer, the patterns specifying the yarns that they sold.

Crompton & Knowles, Loominaries of the Textile Industry

George Crompton (the son of William Crompton) and Lucius Knowles, innovative manufacturers of woolen looms, joined forces to form the Crompton & Knowles Loom Works in 1897. Both companies were based in Worcester, Massachusetts, and engaged in fierce competition for almost two decades, suing each other repeatedly over alleged patent infringements. The competition proved productive, as many improvements were made during this time. But the consolidation was even more advantageous, as with combined resources the company produced the first automatic fancy loom in 1905. Traditional looms had to be stopped every few minutes in order to replace the empty bobbins in the shuttle and this limited the number of looms a weaver could run. Automatic looms have a magazine to hold and replenish weft bobbins while the loom is running. It was said before the invention of the automatic loom that, "A good weaver keeps his

A WEAVER'S WORK
on Ordinary Looms

Operation and Times. (*average approximation*):

Operation	Time
Changing the shuttle	4 seconds.
Replenishing the shuttle (skewering the cop)	11 "
Repairing warp breaks	45 "
Repairing weft breaks	20 "
Finding the pick	2 minutes.
Removing cloth (per cut 80 yds.)	5 "
Cleaning and picking cloth (per cut)	up to ½ hour.
Fetching weft	say 15 minutes per day.
Oiling and sweeping	say ½ hour per week.
Mental anxiety and supervision..	immeasurable.

Advertisement for Northrup automatic loom. This British company made looms very similar to those of Crompton & Knowles.

belt on a tight pulley," meaning that he could change a bobbin without stopping the loom. The result was great variation in the output of weavers, depending on their skill. The invention of the automatic loom meant that a single weaver could keep as many as 8 looms running at a time instead of 2, because they were not required to change bobbins. The automatic fancy loom is truly one of the most complex and ingenious machines to come out of the Industrial Revolution. Its invention was another in the series of revolutionary advances in weaving technology, following the invention of the fly shuttle and pattern chains.

Consolidation also allowed Crompton & Knowles to acquire other specialty loom companies, such as the Gilbert Loom Works in 1899, which made tapestry and carpet looms; the Furbush & Son Machine Company in 1903; and the A.H. Steele Company in 1905, which made narrow fabric looms. The motto of the Crompton & Knowles Loom Works was "a Loom for Every Fabric." The company would eventually manufacture 200 different looms that could weave ribbons $1/8$ of an inch wide to felt looms that wove 50 feet wide. There were looms for wool, cotton, and silk; narrow looms for tapes and ribbons; carpet looms; Jacquard looms; and looms for weaving asbestos brake linings.

The Mills

The Compton & Knowles Company's story parallels developments in the woolen and worsted industry. Companies were consolidating, becoming larger in size and fewer in number. The number of woolen and worsted mills in 1890 was roughly 1,300, and dropped to 750 by 1920. In 1899, William Wood, manager of the Washington Mills in Lawrence, Massachusetts, persuaded seven other mills to consolidate to form the American Woolen Company. Throughout the next 20 years, the American Woolen Company continued to acquire mills, owning 30 by 1909, and 60 by 1920, all in New York and New England. In 1920, American Woolen mills were making 20 percent of the nation's woolen textile and worsted fabric.

WOOL
Golden Age of Industry 1890–1920

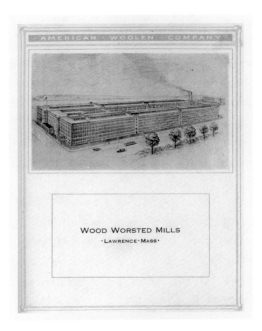

One of the larger American Woolen Company Mills, 1921, from company-published book of same name.

The pioneer mills of the midwest, facing competition from mills in the east with the advent of railroad transportation, were the first to close. Of 183 in 1900, only 70 would still be open in 1920. Some stayed alive, such as Faribault in Minnesota, by finding a specialized niche. In the early 1890s, the Faribault mill made a combination of flannels, cassimeres, and blankets. In 1912, Walter Klemer, grandson of the founder shifted production entirely to blankets, in time to produce 100,000 for World War I. After the war Faribault blankets sold nationally in department stores. J. Capps & Sons, Jacksonville, Illinois, began as a woolen cloth manufacturer, but switched to making Indian blankets beginning in 1892, then converted to war production by 1917. The Flint Woolen Mills of Flint, Michigan, would not be as lucky. They first tried making ready-to-wear clothing from their cloth, switched to carriage cloth in 1904, but missed the boat on automobile cloth when they closed in 1909. Ford's Model T was produced from 1909 and a sample card of their upholstery shows wool broadcloth and Bedford cord swatches.

The type of wool fabric produced by the mills continued to shift from woolens to worsted. Up to 1900, wool fabric produced in the US was primarily woolen. 1909 was the high point of worsteds, when 230 million yards of worsteds versus 89 million yards of woolens were woven. To make worsted, which was a lighter fabric used in suits, both the type of wool used and the preparation for spinning were different. Worsted required longer fibers, which were combed rather than carded. The rise of worsted manufacturing would be aided by the invention of an improved combing machine in 1888. The machinery was more expensive, and was concentrated in the east.

Census data shows a shift to other methods of power from water power: 62 percent of woolen mills were powered by water in 1869; by 1919 26 percent by 1899 were water powered, 43 percent steam powered, and 33 percent electricity powered. More recently established worsted mills reflected their modernity in how they were powered in 1929: 7 percent were water powered, 53 percent were steam powered, and 39 percent were powered by electricity.

The Arts and Crafts Movement

Quixotically, even as woolen and worsted mills muscled up their production and technology, the 1890s brought a return to handspinning and weaving. The Arts and Crafts movement, a handicraft revival inspired by John Ruskin in England, was in part a reaction to industrialization. Ruskin advised that spinning and weaving by hand promise "joy to the maker," in contrast to textile work in a mill. In 1897, an exhibit of handicraft was featured in Copley Hall in Boston. The Boston Society of Arts and Crafts was founded that same year, and published a journal, *Handicraft*, beginning in 1901. By 1907, the Boston Society was one of 33 members of the National League of Handicraft Societies. The Arts and Crafts movement in America conveniently coincided with patriotic societies and a revival of interest in the colonial period and a longing for simpler days. Alice Morse Earle wrote in her

1898 book *Home Life in Colonial Days*, "When a people spin and weave and make their own dress, you have in this very fact the assurance that they are home-bred, home-living, home-loving people." Spinning wheels, handwoven coverlets, and rag rugs became objects of romantic veneration.

Fragment of handwoven overshot coverlet, dog tracks pattern. Mid nineteenth century.
COURTESY OF MACDONALD FAMILY OF SCITUATE, RHODE ISLAND

Wool benefited from the handspinning and weaving revival in the early twentieth century, because of the focus on a uniquely American product. Colonial wool coverlets were once woven from New England to the Appalachians. There were three types of patterns: overshot or "float," another weave structure called summer and winter, and double woven pieces, where the fabric is two intersecting layers. Origins of the patterns could be traced in some cases to European traditions, but not entirely. The distinctively American overshot coverlets had been woven primarily by women, of handspun and vegetable dyed wool on massive hand-hewn looms. Hundreds of new patterns had evolved. These patterns were eagerly collected and old looms resurrected, especially in the Southern Highlands.

Old examples attracted the attention of collectors, notably the Colonial Coverlet Guild of America, founded in Chicago in 1924. In 1913, Mrs. Woodrow Wilson had the furniture in President Wilson's bedroom of the White House upholstered with handwoven fabric made by Elmeda Walker, a Tennessee coverlet weaver.

Wheels turned and weavers tromped handloom treadles in rural and urban settings, from the Appalachian mountains to cities such as Chicago and Boston. Philanthropic societies saw art industry as an alternative to wage labor and factories and as occupational therapy for World War I veterans. Edward Worst of Chicago and Mary Meigs Atwater both started

Blue Mountain Room, White House, 1917.
COURTESY OF THE LIBRARY OF CONGRESS

their teaching careers using weaving as therapy, and went on to write the seminal weaving texts of the time. *Foot-Power Loom Weaving*, was published in 1918 by Edward Worst, who at the time was "Supervisor of Elementary Manual Training and Construction Work" in the Chicago public schools. Atwater worked with World War I veterans in Camp Lewis, Washington, in 1918, but became an authority on handweaving, publishing the *Shuttle Craft Book of American Hand Weaving* in 1933.

Frances Goodrich was another missionary for textile handicraft revival, a social worker for the Women's Board of Home Missions of the Presbyterian Church. After 1895, she experimented with coverlet weaving in the Southern Highlands and organized a guild to purchase materials and market rag rugs and coverlets. She opened Allanstand Cottage Industries, with a showroom in Asheville, North Carolina, in 1908. Other handweaving centers included the Department of Fireside Industries at the Hindman Settlement School, Hindman, Kentucky (1902); the Penland Weavers, Penland, North Carolina (1923); Arrowcraft Shop, Gatlinburg, Tennessee (1925); John C. Campbell Folk School,

Brasstown, North Carolina (1925), and the Churchill Weavers, Berea, Kentucky (1922). From 1902, young women "bartered for learnin" at Berea College, Kentucky, weaving traditional textiles in a log cabin furnished with looms and wheels. The Hull House Labor Museum opened in Chicago in 1900 as a living demonstration of craft processes. It included an exhibit of looms and tools to trace the development of cloth making from colonial to modern looms.

Navajo rug, 1920s.
COLLECTION OF THE AUTHOR

Navajo weavers, settled on the reservation with new herds of sheep, began producing wool rugs and blankets for sale to trading posts. In 1887, a Commission of Indian Affairs agent reported 2,700 blankets made that year, of which two thirds were bartered for goods at the trading posts. The Fred Harvey Company, operators of food concessions for the Santa Fe Railroad, served a function similar to eastern handcraft promoters, by selling rugs at locations along the railroad. In 1900, they established the Indian Building in Albuquerque, New Mexico, where travelers could see Navajo women weaving. Fred Harvey was also responsible for instituting standards of workmanship and design, and helped to set prices. By 1900, rug weaving for a commercial American market replaced Navajo blanket weaving, and traders began to influence design. Rugs were woven in natural colors (red being the exception) and borders were introduced, as in this example at left, bought in the early 1900s by my grandmother.

By 1904, there were more than 50 hand textile industries in New England, the Southern Highlands, and urban areas. Max West was tasked to quantify the activity in a US Department of Labor report. His faint praise was that the textile revival offered "employment for persons living in rural districts and having little else to occupy their time and interest during the winter months, and also for city men and women who are incapable of supporting themselves at more difficult occupations." He references "Indian Work" (basketweaving, beadwork) at the end of his fifty page report, but curiously does not mention Navajo weaving.

The revival of the production of colonial textiles in the early twentieth century was the foundation for later revivals of interest in handspinning and weaving of wool. Dating back to Ruskin, income-generating projects for women, centered around the textile arts of spinning and weaving, would be tried again and again, with a new spin every generation or so.

WOOL'S GAINS and LOSSES:
American Cultural Trends 1920–1950

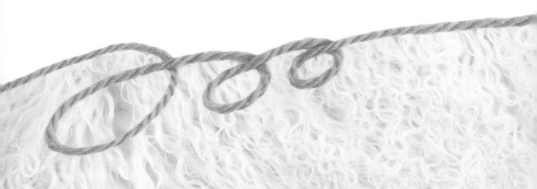

"If clothes were worn solely, or chiefly, for warmth, a union suit and a blanket would answer practically every human need."

Paul Cherington, *Commercial Problems of the Woolen and Worsted Industries*, 1932

In the years between 1920 and 1950, American society shifted seismically. There were changes in the social environment: the nineteenth Amendment in 1920 gave women the right to vote and they became more independent in the workplace and at home as both workers and consumers. Americans' physical environment was also changing, with the electrification of homes, central heating, and the advent of the closed automobile ("America's second home"); and the demographic was changing as Americans continued to move to different parts of the country. In 1900, 62 percent of Americans lived in the Northeast or Midwest. But Americans were increasingly changing climates, which influenced their fashion and fiber preferences. Western states were the most popular, and the South replaced the Northeast as the second fastest growing region by 1930. Finally, changes in the economic environment included the growth of the middle class and the new science of marketing. Marketing experts were trying to puzzle out what motivated Mrs. John Doe and Mrs. Consumer to buy things. These environmental bedrock changes coincided with the development of man-made fibers, beginning with rayon and nylon. The combination would impact Americans' textile needs and choices, wool consumption in particular.

Wool's primary challenge was the radical changes in twentieth-century living environments. Heated apartments, trains, and automobiles meant that people could wear fabrics half the weight of what they had previously worn. A Holyoke, Massachusetts, company, Germania, manufactured wool linings for overcoats. Historian Constance Green of Holyoke, Massachusetts, wrote, "Fundamentally the decline of the company's business was due to the change in fashion effected by the introduction of the closed, heated automobile. Heavy wool overcoatings were doomed." After the precipitous drop from 9 pounds to 3.4 pounds between 1890 and 1920, per capita consumption remained relatively steady, increasing slightly to 3.5 pounds in 1949, with an uptick to 5 pounds during World War II. The market for wool was helped by the fact that even as people wore less wool clothing, it was being used in other applications such as automobile upholstery.

Women's Work

In 1920, women gained the right to vote and began to take advantage of their increased opportunities for life outside the home. Clothing changed accordingly. "One might perhaps say that the discarding of the steel or whalebone corset, the flannel petticoat, in favor of a webbed elastic girdle and cotton or artificial silk underwear, did more for women's emancipation than the vote," said one skeptical but pragmatic observer. At the same time, the era of large homes maintained by domestic help was over. Wages for domestic help went up, and by the Depression it was a luxury. A 1950 retrospective *American Fabrics* article, "A Changing World," described the effect this had. It said, "A silent but ceaseless struggle set in. Women became aware of the cumbersomeness of their families' clothing, and turned to clothes that were easier to wear, easier to work in, and especially those that were easier to launder." Electrification of homes began with the adoption of AC current in 1910 and by 1930, almost 70 percent of homes were electrified. This fueled a home appliance boom. The same article asks, "Who at the turn of the century could foresee the development of the home washing machine to the point where it is almost as common today as the cooking stove?" From 1920 to 1931 the number of electric clothes washers owned grew from one million to over 7 million. In 1933 alone, a million washing machines were sold.

One might wonder which came first, development of the washing machine, or the desire for easy care fabrics? Or did the diminished availability of household help influence both? Other labor saving laundry appliances were selling like hotplates: electric irons and mangles (wringers). Ease of washability was suddenly an important part of advertising and

"Salespeople feel *safer* when they advise Ivory." Ad appearing in 1931 magazines. Warshaw Collection of Business Americana, COURTESY OF ARCHIVES CENTER, NATIONAL MUSEUM OF AMERICAN HISTORY, SMITHSONIAN INSTITUTION

consumer decisions. Wool was a multipurpose fiber long prized for its warmth, strength, and elasticity. However, it was not as well suited to being washed in a washing machine because it felts when exposed to agitation, temperature extremes, and the use of detergents. Ivory Soap was marketed for washing of woolens in machines, but consumers then as now were cautious about washing woolens. A 1928 Ivory ad in *Ladies' Home Journal* tells us "Wool is an animal fiber. It is kept soft by natural oils—just like your hair and your skin. And like your face, wool is sensitive to extremes of temperature, sun, hard rubbing, and not-quite-safe soap. That is why woolens need to be washed with discriminating care, with soap gentle enough for your face." Though wool was durable, the risks of shrinkage and damage by moths began to outweigh its advantages, especially as its primary advantage, warmth, was no longer a critical necessity.

Maytag washer, in *Better Homes and Gardens*.

Fashion and Marketing

Selling Mrs. Consumer, by Christine Frederick, came out in 1929. It is a snapshot of the heady decade of the 1920s, at the dawn of serious marketing efforts directed at women. Frederick writes in the introduction, "Change is in the very air we breathe, and consumer changes are the very bricks out of which we are building our new civilization." There was intense competition for consumer dollars from new goods such as automobiles, washing machines, refrigerators, radios, and vacuum cleaners. As ownership of these goods soared, people were spending proportionally less on clothing. From 1919 to 1930, expenditures on clothing fell from 13.1 percent of all consumer spending to 8.9 percent. One reason was that costs of manufacturing ready-made clothing had dropped, but other critical factors included shorter hemlines and more garments being made of cotton and rayon, less expensive fabrics than wool.

A simple definition of fashion might be "the current style." After World War I, changes in fashion accelerated. As costs of clothing fell, frequent shifts in fashion were more affordable.

Fashion had moved far from the dictates of necessity, and American consumers were influenced by advertising and the mass media.

Copies of clothes from the movies found their way quickly into stores. Magazines such as the *Ladies' Home Journal* and *Good Housekeeping* were full of clothing ads. In 1940, the *Ladies' Home Journal* wrote a year long series, "How America Lives." They noted that clothing had become more informal and better suited to both work environments and also sports activities. Wool, whose features included durability and longevity, was suddenly devalued as fashion shifted more rapidly and people discarded clothes before they were worn out.

In 1932, a brilliant marketing pioneer, Paul Cherington, was among the first to examine consumer's move away from wool in his book, *The Commercial Problems of the Woolen and Worsted Industries*. He appears to have been particularly engaged by wool, as he had authored a book in 1916, *The Wool Industry*, and served as the Secretary Treasurer of the National Association of Wool Manufacturers. He made the connection about the effect of changes in the physical and social environments on clothing choice. Cherington notes, "The winter street costume worn by most women in 1921 and in 1931 will indicate the increasing lightness of equipment:

>1921: Union suit, Bone corset, Camisole, bloomers, petticoat, heavy lined wool dress

>1931: All in one (girdle and chemise combination), slip, light wool dress"

Until 1920 people still wore union suits, even though central heating with steam and hot water radiators had been in use in parts of the country for 50 years.

Cherington also noted that the United States was not a single market but was influenced by region, ethnicity, and economic status. Also, the location of the market was a moving target. Wool might sell briskly in New England and other northern climes, but increasingly Americans were moving south and west.

Ernest Dichter was a later marketing pioneer, a German with a doctorate in psychology who arrived in the US in 1938. Intuiting that the middle class tended to rely on advertising and mass media for tips on what to buy, he started a consulting firm, Institute for Motivational Research, to find out what motivated consumers. He went on to create memorable enticements such as the famous "tiger in the tank" gasoline ad and the "bet you can't eat just one" potato chip ad.

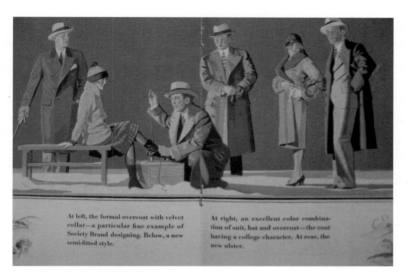

"What Men Wear Everywhere." 1926 catalog for Society Brand Clothes.
Warshaw Collection of Business Americana,
COURTESY OF ARCHIVES CENTER, NATIONAL MUSEUM OF AMERICAN HISTORY,
SMITHSONIAN INSTITUTION

The 3-piece suit was still a man's uniform, making men's apparel a large part of the market for wool fabric. However, fashion changes affected menswear also, as lightweight suits replaced the traditional blue serge. Cultural changes affected menswear, as when the 1938 federal law created the 40-hour workweek, an increase in leisure time called for leisure or casual wear choices. The California clothing industry grew, promoting its California casual lifestyle look, with lightweight suits of cotton and rayon replacing wool suits.

New Competition

Wool had always competed with other textile fibers, especially cotton. Cotton is a general purpose fiber. It is less expensive than wool by anywhere from one third to one sixth, depending on wool prices. It was primarily manufactured as "grey goods," meaning it was woven as plain white cloth and was stockpiled until needed, when it would be printed, dyed, or otherwise finished. Wool, in contrast, was typically manufactured for a specific purpose, with color and design woven in, and no options for modifying it if styles changed.

However, interfiber competition intensified at the beginning of the twentieth century with the development of man-made fibers. In 1887, "artificial silk" fibers developed by Europeans were exhibited at the Paris International Exhibition. By 1911, the new fiber was in production at a plant in Pennsylvania. In 1925, American factories produced 51 million pounds of the fiber now named rayon. Production rose to 270 million pounds in 1935, or 10 percent of all fiber used in apparel. Rayon was a manufactured cellulosic fiber, and its price dropped quickly. In 1930 it was more expensive than wool ($1.05 per lb. to wool's $0.63 per lb.) but just four years later the values had reversed, as rayon fell to $0.58.

From the beginning, the man-made fiber industry benefitted from substantial research and development funding, and large amounts of advertising. DuPont, which had originally made its fortune producing gunpowder, opened a viscose rayon plant in 1927. In 1928, DuPont began research into the "how and why of molecules," attempting to

Rayon dress ad, circa 1950.

discover how molecules unite to form giant molecules such as cotton and silk. It was this research that eventually led to the development of nylon, the first synthetic fiber. DuPont applied for a patent in 1938, giving it a name in the generic category, like rayon. It was a strong but elastic fiber, and was lauded for being "a synthetic symphony from DuPont's pile of coal." DuPont implied nylon's simple and natural origins, advertising it as being made from air, water, and coal.

Corporations such as DuPont and Monsanto operated on a much different scale than traditional textile industries, and had the resources to develop new products and promote them. Later, they would sell them by the marketing practice of branding, but for the moment nylon and rayon joined the fiber scene in the same generic category as wool, cotton and linen.

Large crowd lining up for nylons in San Francisco, 1946.
IMAGE COURTESY OF THE HAGLEY MUSEUM AND LIBRARY

The first consumer products made of nylon, stockings and toothbrushes, were unveiled at DuPont's Wonderful World of Chemistry pavilion at the New York World's Fair and the Golden Gate Exhibition in 1939. Stockings in particular were an instant success, though not available for civilian consumption because in 1942 the War Production Board had commandeered all nylon production for parachutes, ropes, tents, uniforms, and other war needs. Nylon stockings were not available again until after the war, when the media reported nylon riots, with hundreds of thousands of people lining up to buy them. Per capita consumption of man-made fibers rayon and nylon increased to 9.5 pounds by 1950 from a dead start of zero in 1920.

New Markets

Despite the competition with synthetics, during the decades of 1920 through 1950 wool found some new markets. After World War I, interest in sports and outdoor activities exploded. Men and women bicycled, played tennis and golf, and went to beaches. The original tennis whites were wool, as were bathing suits. Many lightweight sports costumes were fashioned from the original miracle fiber, wool, because of its elasticity when knit. The knitting industry expanded into bathing suits in the 1920s, and later, as "casual" clothes became popular, into fashion sweaters.

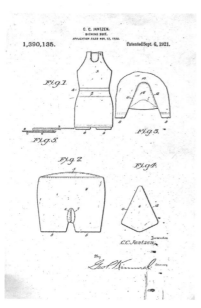

Jantzen patent. Warshaw Collection of Business Americana,
COURTESY OF ARCHIVES CENTER, NATIONAL MUSEUM OF AMERICAN HISTORY, SMITHSONIAN INSTITUTION

The Jantzen Company slogan was "The Suit that Changed Bathing to Swimming." To this end, they applied for and received a patent in 1921 that stated, "The primary object of the invention is directed to a novel design of bathing suit which will not bind the body and permits freedom of movement." Another bathing suit manufacturer, Catalina (formerly the Pacific Knitting Mill), was founded in 1928 and became famous for their "Styled for the Stars" Hollywood label. Speedo introduced the first non-wool suit in 1928, but swimsuit makers continued to use wool through the 1930s. By the late 1930s and '40s there was finally a shift to using cotton and rayon; these were less expensive materials. Also, as standards of modesty relaxed, fashion could play a greater role in swimsuit design.

LEFT:
Ford closed car brochure, 1920s.
COURTESY OF HAGLEY MUSEUM LIBRARY

RIGHT:
Ford V-8 ad, 1932.
COURTESY OF HAGLEY MUSEUM LIBRARY

Automobile interiors were another new market for wool. "Closed" automobiles began to be manufactured in the early 1920s. Car ownership was a rapidly growing market. In 1910 there was a car for every 43 families, by 1920 one car per 4 families, and by 1930 more than 1 in 2 families owned automobiles. It was a significant market for wool cloth because the early automobiles were upholstered with traditional wool upholstery fabrics, an updated version of carriage cloth. In 1922, a survey of car owners found that 75 percent of car owners preferred (wool) fabric to leather. "A well executed automobile interior has become a unique expression of our time and spirit and a thing of beauty." Designers, recognizing that automobiles were an extension of living space, applied the same standards of décor and catered to the same clientele. "He buys the car, she buys the upholstery" read one ad. In a 1940 study, Ernest Dichter wrote, "Women use the same buying criteria in judging cars as they use in two fields with which they are most familiar . . . fashion and housekeeping. If details she can judge from her experience are of good quality . . . such as upholstery, color finishes and so forth, she concludes that the entire car is of good quality."

Ford upholstery, 1930s.
COURTESY OF HAGLEY MUSEUM LIBRARY

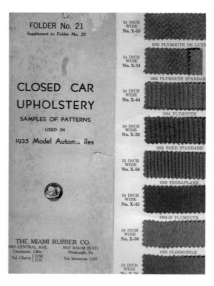

Miami Rubber Company, samples of patterns used in 1935 automobiles.

Fortune magazine ad, 1933.
COURTESY OF HAGLEY MUSEUM LIBRARY

The Census of Manufactures reported in 1935 that 7.5 percent of wool cloth manufactured was for automobile cloth. The American Rubber Company of Cincinnati's 1935 sample card shows wool swatches for "Closed Car Upholstery" for 8 car companies: Buick, Chevrolet, Dodge, Ford, Oldsmobile, Plymouth, and Pontiac. Car buyers could choose custom upholstery, upgrading from standard to deluxe. Some wool mills such as the Yale Woolen Mill, only 54 miles from Detroit, Michigan, switched from apparel production entirely to making wool broadcloth for car interiors. They expanded from 36 to 112 looms in the 1920s, and bought 76 new Crompton & Knowles woolen looms during the period 1934 through 1936. Ford and General Motors both opened textile divisions. By 1940, the American Sheep Industry Association reported that 12.5 percent of wool cloth production was for use in automobile interiors.

Chatham blanket ad featuring opera star Lily Pons, *Ladies Home Journal*, September 1932.

Industrial map of the United States, 1921.
RAND MCNALLY

After making blankets for a half century, Chatham Blankets began making automobile upholstery in 1936. They would also manufacture electric blankets.

The midpoint of this period, 1937, saw two distinct categories of wool growers: forty-two percent being territory wool growers with large bands of sheep on open ranges, and thirty percent being farm method (fenced) wool growers, with small flocks producing less than 350 pounds a year. The remaining growers were from Texas, practicing something in between as they fenced flocks on very large ranches.

Politics and Wool

During the Depression, there was an increased use of cotton and rayon for apparel, as they were less expensive than wool. But even before the Depression began, the wool industry sustained major losses. Consumption of wool dropped from 312.8 million pounds in 1922 to 167 million in 1934. The government responded with efforts to support the wool industry. The Buy America Act was passed in 1933, requiring manufacturers supplying the US Army to use domestic wools if available. In 1935 the government purchased and stockpiled huge amounts of wool goods, somewhere between 35 and 50 million pounds. The Wool Products Labeling Act, effective 1939, required anyone who manufactured or sold products containing wool to accurately label each item with the fiber content and origin. This was in part a recognition that the identification of fibers, previously an easy job for consumers, had become more complicated with the development of man-made fibers.

Army specifications for overcoating, 1942.

As with previous wars, World War II gave an enormous boost to the woolen and worsted industry with the need for uniforms and blankets. From 1942, civilian production of woolens ceased, as all efforts were directed towards war needs. A Crompton & Knowles ad notes that "it takes 10 sheep to clothe one soldier." Mills rolled out wool fabric for everything from underwear to blankets. The Beacon Blanket Company of Swannanoa, North Carolina, known for its elaborately patterned Jacquard woven cotton blankets, converted to wool during World War II, producing 7 million blankets. In 1942 the Cavendish mill of Vermont made 200,000 blankets, plus overcoating. Even the American Woolen Company came back from the dead to make profits of $88,000,000 between 1939 and 1948.

Industry Economics

A Wharton Business School professor and industry observer wrote in 1935, "Instead of goods selling themselves, now they can be sold only after intense effort in market analysis, merchandising, sales administration, and advertising." During this period, the economic horizon tilted away from the textile industry, as cloth production changed from being a seller's market to a buyer's market. From the beginning of the Industrial Revolution through the end of the nineteenth century, manufacturers had control over what they made and who they sold it to. After 1920, the ready-to-wear clothing industry was in a position to dictate to the mills what they wanted to buy. There was an unfortunate lack of coordination between what they wanted to buy, and what was being produced. Larger mills, American Woolen in particular, were disadvantaged in the fast changing industry, as they were less able to respond to market trends.

Fashion advisory services offered advice to mills and merchants trying to keep pace with fashion trends. Tobe Davis was perhaps the first professional fashion stylist, pronouncing on shape, design, and especially color of consumer goods from apparel to automobiles.

Wool's Gains and Losses: American Cultural Trends 1920–1950

In 1935, wool manufacturers, dealers, and growers organized the Associated Wool Industries to promote the use of wool, following a wool consumption low in 1934. Their activities included informing consumers of the value of wool and aiding retailers towards more profitable selling of wool merchandise. This organization was disbanded after only 4 years. Wool growers then turned to the International Wool Secretariat, and formed the American Wool Council in 1941.

In the mills themselves, operations were mostly unchanged since the invention of the automatic loom in 1905. The machinery was old but still functional, which encouraged production of traditional fabrics. Shifts were made to produce cloth for automobile manufacturers, but it was a lateral shift rather than innovation of a new type of cloth. Crompton & Knowles, the leading manufacturer of woolen and worsted looms, moved into making equipment for industrial applications, such as felt for the papermaking industry. In 1935 they commissioned a survey of mill equipment, *C&K Box Loom Census*. The report published in 1937 noted that 50 percent of looms were obsolete, or over 25 years old. Development of new technology was stalled during the Depression, as mills had more than enough production capability.

In 1939, the Census of Manufactures ranked the woolen and worsted manufacturers seventh in the number of wage earners employed. Change in the industry can be tracked by looking at the number of mills, and who owned them. In 1914, there were 1,070 woolen mills. Consolidation in the industry began with American Woolen Company acquiring 60 mills in eight states by 1924. However with

Crompton & Knowles loom for weaving felt for papermaking, circa 1940.
COURTESY OF THE AMERICAN TEXTILE HISTORY MUSEUM

Weave room, Charlton Woolen Mill, circa 1935.
COURTESY OF THE ALIX FAMILY

the sharp decline in demand for blue serge in 1923, the company lost its staple product. William Wood's successor, Andrew Pierce, sold three mills and closed 27 by 1926. In 1929, the Census of Manufactures reported 724 mills, then 600 in 1939. The 1947 census shows 495 woolen and worsted mills, with 399 in New England and the Middle Atlantic states. Half of these had fewer than 500 employees, and one can picture the small woolen mills of New England in these figures.

Responding to Change

The military needs of World War II could have been an incentive for the wool industry to innovate. Unfortunately, any innovations had nothing to do with wool. In its 1944 Annual Report, Crompton & Knowles predicted, "Our first new product to be made when conditions permit will be the S-6 Rayon Loom." After World War II, there was an explosion of pent up demand for consumer goods. Initially wool fabrics and the

Crompton & Knowles brochure, 1950s.
WORCESTER PUBLIC LIBRARY

machinery to make it sold well. The 1948 Census of Manufactures reported that in 1947, $50 million was spent on new plants and equipment in the woolen and worsted industry. Some of this money went to buy Swiss-made Sulzer shuttleless looms, replacing Crompton & Knowles in woolen equipment manufacture.

By any measure, the pace of cultural change during the period 1920 to 1950 was frantic. New products, household technology, and cultural norms entered the marketplace. Consumer familiarity with wool allowed it to compete at least initially for new markets. Eventually, however, the woolen industry's relative sluggishness to imagine and anticipate new demands, and its displacement by man-made fibers would result in its decline as an important manufacturing industry. There were no developments in wool textile technology comparable to what was happening with man-made fibers and technology. There would be lasting effects on consumers' reliance on wool as the original all purpose fiber.

LOSING GROUND
Post–World War II 1950–1980

"Presented commercially in 1950 and 1951, these manmade fibers offer such great promise that within a decade the sheep may indeed have their wooly backs to the wall."

Paul Mazur, *The Standards We Raise*, 1953

Losing Ground: Post–World War II 1950–1980

The 1950s were the pivot point for the wool fiber's commercial viability. In 1952, Alec Guinness starred in a British film, *The Man in the White Suit*, playing an inventor who dreams of creating an indestructible fabric that can't be torn, frayed, or stained, an invention which would bring about the extinction of the textile industry. The inventor is up against a textile mill owner making traditional men's wool suiting. Although Guinness's character loses in the movie, history vindicates him. The movie speaks to the challenges of the post-war textile industry, with man-made fibers and new fabrics replacing the old ones, wool in particular.

The population of the US, 152 million in 1950, added an average of 25 million new bodies to be clothed each decade through 1980. Initially there was a pent-up demand for wool which had largely been commandeered for military needs during the war. But in short order, wool was replaced by man-made fibers, wash and wear fabrics, and even new constructions such as double knits, all invented in the decade following the war. Domestic consumption of man-made fibers increased 73 percent from 1949 through 1969, with the result that per capita wool consumption was only 1.5 lbs. by 1969. Innovation in traditional industrial textile technology ceased, textile production moved overseas, and American mills began to close. Crompton & Knowles, makers of virtually all woolen and worsted looms used in the US, dropped Loom Works from their name in 1954 and primarily sold parts for their looms rather than manufacturing whole machines. But at the same time, the 1950s saw the beginning of another handweaving revival, this time with "modern" handlooms made by dozens of small companies.

Man-made Fibers

> "Only a brash person would forecast the complete
> supplanting of wool by synthetic fibers, but it would
> also be foolish for the wool manufacturer to underestimate the
> competitive strength of the newer synthetics."
> JAMES MORRIS, 1952

DuPont had invented nylon, and continued to be exceptionally successful in the development of new fibers with the introduction of trademarked fibers: Dacron, a polyester, and Orlon, an acrylic. Both were low maintenance, being easily machine washable and dryable. Dacron's wicking qualities improved on nylon's, and in addition could be permanently creased or pleated. In 1951, Dacron made a splash with the story of "the swimming pool suit." This was a man's business suit, 100 percent Dacron, worn continuously for 67 days and then dunked twice in a swimming pool. It was washed once in a washing machine, and without ironing reportedly looked presentable when shown to the New York press.

Swimming pool suit, 1950s. From *American Fabrics* magazine.

The Textile Fiber Products Identification Act, effective in 1960, established 16 generic terms for man-made fibers (acrylic, modacrylic, polyester, rayon, acetate, saran, vinyl, olefin, vinyon, azlon, nytril, nylon, metallic, glass, rubber, and spandex) in addition to copyrighted fabrics such as Orlon, Dacron, and Acrilan. By 1960, 29 percent of fiber used by mills was man-made, 65 percent was cotton, and 6 percent was wool.

American Fabrics Magazine and Wool Marketing

The first issue of *American Fabrics Magazine*, a quarterly textile trade periodical, came out in 1947. It mixed advertisements, informational articles about new fibers, inspirational articles about art, designers, and traditional ethnic textiles, with company news. Actual fabric swatches were pasted into both advertisements and the section on trends. It was read by clothing manufacturers, department store executives, manufacturers, and designers. A person paging through old issues will be thoroughly enchanted by the fashions and attitudes of those days. There were provocative ads, clearly relishing the opportunity to picture women's underwear and nightclothes. There were ads for formal wear, photographed in highly scripted situations. Wool garments featured prominently in early issues, with many ads and articles. The magazine's editors demonstrated their commitment to wool by devoting a large part of the 1948 issues to a "Dictionary of Wool and Worsted Terms." However, the story of textile consumption through the years of the magazine's publication (ending in the 1980s) is clearly told. Here is a quote from issue number 21, in 1952:

"Until a few years ago fabric designers were restricted by the limitations of natural fibers . . . . Because these (man-made) are engineered fibers, they can be endowed at their birth with many unique functional properties that can be translated by the genius of designers into fabrics of *hitherto unknown beauty, durability, and economy*." An issue 12 years later noted that the US Summer Olympic team in 1964 was outfitted in uniforms made largely of synthetics.

Ernest Dichter was hired to speak to the National Association of Wool Manufacturers in 1956. In his address, "Tear the Grey Cloak," he told them, "Wool and wool garments are being created, designed,

Encron ad, *American Fabrics* magazine,
Summer 1968, issue 79.

manufactured, and merchandised in ways more reminiscent of the days of Victoria than of the Jet." In the late 1950s Dichter addressed the need for fanning the flames of ardor for textiles by explaining the emotional appeals of wool, cotton, and silk. With so many new choices, textile consumers needed to be wooed. Dr. Dichter made emotional associations between each of the natural fibers: cotton was soft, friendly, clean, cool; silk was feminine, intimate, refined, and sensual; wool was masculine and tough, sedate, respectable, conservative. Wool was a protector, and promoted group solidarity. In a 1969 speech to a textile marketing forum, *Love the Fiber—How to Motivate the Textile Consumer,* he would say "A fundamental mistake is for an industry to take itself for granted and to assume that all it has to do is to produce good products and the rest will take care of itself. The modern consumer is being wooed from all sides." Ultimately the wool industry, including major equipment manufacturers such as the Crompton & Knowles company, was not able to keep pace with these changing consumer needs.

Estelle Ellis, another marketing pioneer, best known for cofounding *Seventeen* magazine in 1942, astutely connected fashion with changes in society. She observed, "Fashion is about change and changes in materials are changing the way we live. Materials define our age historically. Now materials are designed to fit specific needs."

Trends in Consumption

Consumers had more choices in both fibers and labor saving appliances. Their use of automatic appliances had an effect on fiber consumption. A half million electric clothes dryers were sold in the 4 years after the war. A few years later, a 1958 high school textbook encouraged students to conduct their own time and motion study on savings in time and energy using a dryer versus drying laundry outdoors. The message was clear that a dryer was a great time saver for homemakers. How did this affect wool? Wool cannot be dried in an electric dryer, as without extreme caution it will felt.

Sensing the change in demand for wool, in 1957 the US Department of Agriculture (USDA) commissioned a series of marketing research reports: Number 152, "Fabrics and Fibers for Passenger Cars: Automobile Manufacturers' Views, 1955 Compared with 1950"; Number 153, "Women's Attitudes towards Wool and Other Fibers"; and Number 155, "Teenage Girls Discuss Their Wardrobes and Their Attitudes towards Cotton and Other Fibers." These would be followed by the 1959 survey Number 338, "Consumer Concepts of Fabric: A Marketing Survey of the Relative Importance of Fabric Characteristics in the Selection of Women's Clothing," and in 1963 Technical Bulletin No. 1301, "Demand for Textile Fibers in the United States."

Ford 6 for 1950 brochure, featuring wool broadcloth upholstery.

With respect to the first marketing research report, the horse was already out of the barn, or the car out of the garage, as nylon found its way into car interiors in the early 1950s. The USDA marketing research report Number 152 noted that from 1950 to 1955, wool went from being 51 percent of material used in car interiors to 1 percent. Automobile executives were surveyed about characteristics of upholstery materials: performance, appearance, trends in style. They were asked, "What changes do you expect in the fiber content of upholstery fabrics used in future cars?" Sixty-nine percent predicted an expectation of more synthetics, nylon in particular. One executive commented, "More nylon content in blends—between 30 and 50 percent of total weight. This will give us longer life, strong wearing qualities, appearance, luster, and a name we can merchandise." Wool had lost an important market.

For the second marketing research report, "Women's Attitudes Towards Wool and Other Fibers," the USDA Market Development Branch conducted personal interviews of a representative sampling of homemakers in their houses. "At the time of this national study in late

WOOL
Losing Ground: Post–World War II 1950–1980

Ford brochure, 1951, featuring Craftweave upholstery, "an unusually durable material of combined nylon, rayon, and cotton."

Ford Sunliner brochure, 1952, featuring vinyl upholstery.

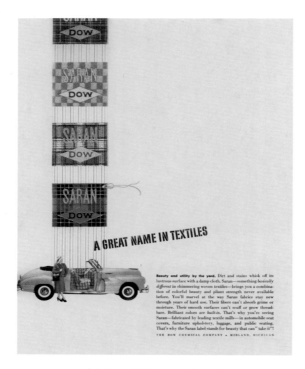

American Fabrics magazine ad.
Dow Chemical Company

Rayon upholstery with coordinating rayon dress,
American Enka Corporation, *American Fabrics* magazine, 1955.

1955, wool was the primary fiber used in American women's fall, winter, and spring suits and skirts." Sixty-three percent of women interviewed reported that they owned at least one suit. Of those who owned suits, the average owned 2.6 suits, three quarters of suit owners reporting that they owned wool suits. They rated wool highly: wears well, holds its shape, doesn't wrinkle easily, warmth. The response of sweater owners was a little different, in that more women owned Orlon sweaters, preferring them for their washability and shape retention. Teenage girls were the largest share of the sweater market, owning an average of 7.4 sweaters each, as many Orlon as wool. Nine out of ten women interviewed said that they were aware of the fiber content of what they were buying, and used that knowledge in their decision to purchase.

US Department of Agriculture
Marketing Study Report.

Knit wool uniforms, 1968.

In reporting on how they cared for their garments, 7 out of 10 had suits dry-cleaned, while 3 out of 5 washed their sweaters. Most women who owned wool items were aware of the need to protect wool from moths, and stored wool appropriately, either in garment bags or cedar chests.

Mothproofing device for wool clothes, twentieth century. PHOTO BY JOANNE SEMANIE

Stevens and Sons Company,
American Fabrics magazine, 1948.

Part II of the study interviewed women about their sewing, knitting, and needlework. The report found that "the majority of women in the United States perform some type of needlework," 7 out of 10 sewing with a machine, 1 out of 10 knitting. Ninety-seven percent of sewers sewed with cotton, 30 percent with wool.

Miron Mills, *American Fabrics* magazine, 1947.

Draper Woolens, *American Fabrics* magazine, 1948.

Losing Ground: Post–World War II 1950–1980

Wool branding had been attempted by woolen mills after the war, advertising brand names such as Botany, Draper, Forstmann, Farnsworth, Hockanum, Miron, and Somersville. Although these woolen companies were churning out spring and fall lines of wool fabric every year, the 1955 study discovered that 43 percent of the women surveyed would not wear wool dresses and suits.

The top three advantages noted by women about synthetics were the ability of the fiber to hold its shape (nonsag), wrinkle resistance, and color fastness. Orlon replaced wool as the predominant fiber in sweaters, especially among the younger generation. The American Wool Council mounted a campaign to educate consumers, and glamorize and promote wool.

Susquehanna Mills, *American Fabrics* magazine, 1948.

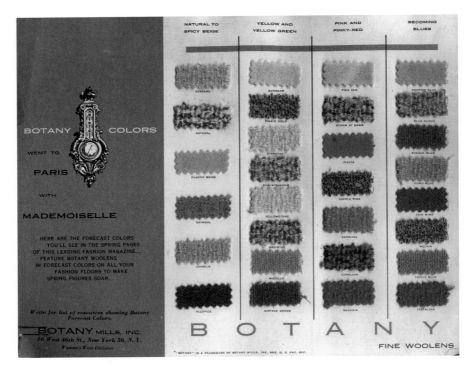

Botany Mills Spring 1953 line of woolens and worsteds.

Wool used in blankets declined from 30 percent in 1949 to less than 7 percent in 1959. Electric blankets had been invented after the war and patented by General Electric (GE), and woven blankets were increasingly being made of acrylic and polyester. A Woolrich executive observed wryly, "Who needs a blanket—just turn up the heat."

Losing Ground: Post–World War II 1950–1980

Tomorrow's Fabrics

In 1962, an issue of *Textile World* exhorted manufacturers, "If you're not using man-made fibers today, now is the time to get in on the ground floor. You can gear production, fabric developments, and new equipment purchases you may be considering for tomorrow's fabrics . . . Don't sit idly by, or you'll run the risk of being second best in an important and expanding market."

New fibers were developed into new types of fabric: wash-and-wear, bonded fabrics, and knits. Wash-and-wear fabrics offered freedom from household tasks in the late '50s, bonding stabilized wovens and knits and offered both insulation and self lining. Knits as an industrially manufactured

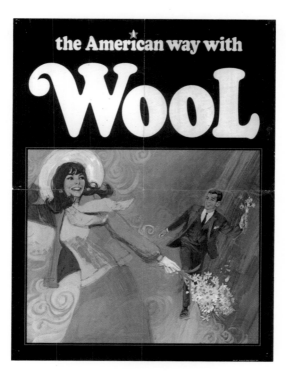

American Wool Council poster, circa 1970s.
COURTESY OF THE AMERICAN SHEEP INDUSTRIES ASSOCIATION

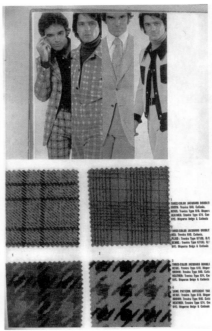

LEFT:
Deansgate man's suit, wool, 1956–1959.
COURTESY OF ARCHIVES CENTER,
NATIONAL MUSEUM OF AMERICAN HISTORY, SMITHSONIAN INSTITUTION

RIGHT:
Trevira doubleknit fabric and suits, from "A Major Advance by
Hoechst Fibers," American Fabrics Marketing Report, 1972, issue 94.

textile structure had been around since the patenting of the latch needle in Britain in 1849. However as clothing fashions continued to evolve, by 1968 knits represented 25 percent of overall textile production. The year of the men's double knit suit was 1968. Knits continued to grow in popularity as consumers embraced informal activewear: t-shirts, sweatshirts, and sweatpants. To confirm the widespread consumption of knitted fabrics, take a quick inventory of your own closet. There is a good chance that much of your clothing is knitted rather than woven.

Presciently, a 1968 *American Fabrics* article said:

> There is no doubt that knits are where the action is. One respected industry spokesman, for example, told us a few days ago that the head of a major mill has said that he has bought his last loom—as he calmly installed a battery of knitting machines . . . [The article goes on to say] In the first place, the knit is a perfect reflection of our new set of moral values. It is eternally sexy, sensual, full of texture and seductiveness . . . In the second place, it is the ultimate answer to the second skin.

Eventually easy care double knits made inroads into men's suiting, where wool was already feeling competition from synthetic fibers.

Changes in Technology

In 1949 John Tinsley, president of Crompton & Knowles, gave an address, "Looms for the World" to the Newcomen Society in North America, an organization studying the history of engineering and technology. Tinsley stated, "The position of the loom today is secure through its accomplishments down the ages in peace and war. It is perhaps second in importance to no machine devised by man. In its most significant aspects—the benefits it has bestowed upon Mankind—it ranks with the plow, the printing press, and the locomotive." As technology has progressed, this list is positively quaint, but his remarks demonstrated the blind spot in the road that Crompton & Knowles would travel.

The workhorse of the woolen and worsted industry, the Crompton & Knowles W-3 loom, was outmoded in the 1950s by the Sulzer loom, a shuttleless projectile loom that wove one and a half times faster. The Sulzer loom, developed in Switzerland, used a projectile to move the weft, a light clip that worked by grasping the weft end from a cone of yarn on one side of the loom, carrying it across and then dropping it off at the other side to be conveyed back to the original side to do it again. This replaced the traditional bobbin and shuttle and ran at faster speeds, approaching 230

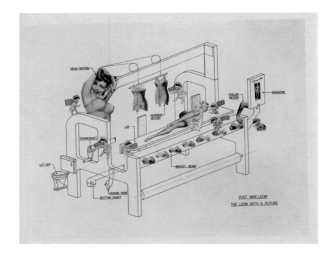

Loom builder's fantasy loom, Crompton & Knowles Records, 1940s.
COURTESY OF THE AMERICAN TEXTILE HISTORY MUSEUM

BELOW:
Multi-purpose loom, from Crompton & Knowles Annual Report, 1955.

picks per minute as compared to the W-3's 180 picks per minute. After World War II, the Crompton & Knowles's annual company reports tell a startling story. In 1954, Loom Works was dropped from the company name, and it became the Crompton & Knowles Corporation. From that moment, the company pursued diversification in the manufacturing of dyestuffs, automobile parts, and even an automatic pin setter for bowling. The company came out with a multipurpose (MP) loom in

1955, suggesting that MP could mean "more profits." In 1959, they developed an electric push button loom, the C-7, so operators with little experience could run the looms. This addressed the fear that looms were "complicated machines that have reacted violently when run by an inexperienced person." Until now, weavers were among the most highly skilled workers in a mill.

When Crompton & Knowles ceased loom production, man-made fibers dominated the textile industry. The technology of looms was changing, partly in response to the different requirements of man-made fibers. Beginning in the 1950s, the goal of loom manufacturers was to increase the speed of weft inserting and to reduce noise. Shuttleless looms were developed to replace the conventional shuttle loom. They operate at greater speed, anywhere from the 230 picks per minute of Sulzer looms to 1,250 picks per minute of air jet looms, because the shed does not need to open so wide as to admit a shuttle and a heavy shuttle does not need to be thrown. In addition, the task of winding bobbins is eliminated. Modern looms use compressed air, water, darts, or rapiers to propel the weft through the warp. All of these machines work by grasping the yarn end from a cone set on one side, carrying it across the warp, cutting it at the edge of the cloth, and starting again. Cloth made in this way will thus not have a woven edge, as each pick is laid in separately. Rapier looms were the first type of projectile mechanisms invented. Instead of a shuttle, they use a rapier device (for rapier, think long slender sword) to travel the width of the cloth to carry the weft. An early version of a rapier loom was developed to weave horsehair.

Significantly, none of the companies manufacturing modern looms were American. New weaving innovators included Northrop (Britain), Sulzer and Saurer (Switzerland), Picanol (Belgium), Dornier (Germany), Vamatex (Italy), and Tsudakoma, Suzuki, and Toyota (Japan). Originally founded by Sakichi Toyoda in 1926 as Toyoda Automatic Loom Works, Toyota would switch gears (so to speak) to become more famous for automobiles.

Ralph Huey, a retired Crompton & Knowles salesman wrote pensively in a 1985 letter, "I don't believe that many of us long-timers ever envisioned the fate that befell C&K. The foreign firms learned to respond quickly to USA mill demands and they really came up with some wonderful new ideas and technology."

In 1979, the Crompton & Knowles Corporation closed its Worcester plant, and mothballed the buildings. The 1978 Annual Report noted, "One major business of C&K which failed to perform to expectations in 1978 was Textile Machinery." The company had been developing its first shuttleless, or projectile loom, the 400 Series. The *Worcester Herald* noted of this loom, "It was all but the last gasp in the textile machinery business." It was 7 years in development, with several million dollars invested. A New England company, Hayward-Schuster Woolen Mills of East Douglas, Massachusetts, ordered 24. However by 1981 the two companies were in court, with Hayward-Schuster asking for $3.5 million in damages, saying the loom hadn't performed as advertised. With this, Crompton & Knowles' loom building history was effectively over.

Politics and Wool

Congress passed the National Wool Act in 1954, in an attempt to support wool production as "an essential and strategic commodity." In 1955, the National Association of Wool Manufacturers reminded the US Department of Commerce that "An atomic war will probably put a still greater strain on production facilities for warm clothing and cover. The manufacture of wool tops, yarns, cloths, and blankets in this country must, therefore, be maintained in the national interest." The National Wool Act increased tariffs on wool and wool products and established an incentive payment program which established a target price for raw wool. However, numbers of sheep continued to decline. The program was phased out in 1995, and the act repealed on the recommendation of the US Department of Agriculture. The reason given was that wool was no longer considered a strategic material, having been removed from the Pentagon's Defense Logistics Agency Strategic Materials list in 1960.

By the 1960s, manufacturing was moving overseas where wages were 3 to 12 times lower than those paid in the US. In 1969, 2 out of every 5 men's suits sold in the US were made of cloth woven in Japan. As mills closed and imports of woven fabrics soared, used machinery dealers scavenged the New England mills, exporting liquidated equipment to developing foreign nations.

Freshman Senator John Kennedy wrote an article for the *Atlantic Monthly* in January 1954 to explain a bill he had introduced in 1953 to expand and diversify industry in New England to replace the lost textile industry. Massachusetts, the first state to industrialize would also be the first to deindustrialize. As Massachusetts had led the country both in woolen production and in manufacturing textile equipment, it was the hardest hit by the loss of this traditional industry. One example: after losing millions of dollars, in 1955 American Woolen, the company owning 25 percent of the New England woolen mills, merged with Textron, Inc., who had begun in 1923 as a rayon enterprise. Textron liquidated all its New England mills by 1957. The Yale Woolen Mill in Michigan, which had done so well with automobile fabrics, converted to wartime production but was unable to reenter the automobile business after the war, as carmakers switched to synthetics. The Yale Mill closed in 1963.

But the national woolen industry was not dead, as 100 new woolen and worsted mills opened in southern states in the early 1950s, lured by the lower costs of manufacturing. Elsewhere in the country, the Reedsburg Woolen Mill in Reedsburg, Wisconsin, survived into the 1960s with novelty yarns and fabrics, its economic viability based on the talents of its designers.

The National Association of Wool Manufacturers consolidated into the American Textile Manufacturers in 1971. This organization in turn dissolved in 2004.

The Handcraft Revival

Paradoxically, as labor-saving home appliances proliferated, a revival of labor intensive textile production was occurring. New handspinning and weaving equipment was being manufactured, and new books and periodicals published with information about how to use them. In the first issue of one of these, *Handweaver & Craftsman* (published from 1950 through 1975), editor Mary Alice Smith asked, "Why a magazine for handweavers? For those who haven't been following the exciting progress of weaving and other important handicrafts in the United States, the project at first glance might seem like a sort of silly business. Handweaving, they tell us, disappeared with the ox cart. That is just their mistake. There were, a sound trade source estimated about two years ago, some 125,000 handweavers in the United States and Canada."[237]

Regular contributors to *Handweaver & Craftsman* included Berta Frey, Mary Atwater, Lili Blumenau, and Harriet Tidball, all authors of books on handweaving during this period. Another regular, Walter Hausner wrote a series of articles about man-made yarns, "Characteristics of Man-made Yarns"(1955), and "Developments in Man-Made Fibers" (1960).

Unlike the colonial revival weavers at the turn of the century, handweavers of the 1950s and '60s were experimenting, choosing both traditional and nontraditional designs and materials. The second issue of *Handweaver & Craftsman* would feature an article about yarns available to handweavers. Traditional yarns of wool and cotton were staples, but handweavers were also open to new fibers: rayon, nylon, and metallics (including Lurex, a metallic yarn made of a laminate of aluminum foil and transparent film, sliced into strips). They used newly engineered handlooms that borrowed from industrial technology. Because they had grown up with the ethic of Depression era frugality, they also might use such surprising resources as old nylon hose, vegetable bags, and recycled wiring from electric blankets (all as recorded in *Handweaver & Craftsman* articles).

The revival of handspinning coincided with the back-to-the-land

Looms for TV Queens

Certain participants on the popular *Queen for a Day* TV and radio programs now own handlooms and a variety of loom accessories. The presentation of these prizes has been arranged by the Grant Handweaving Supply Co. of Los Angeles with the cooperation of the Structo and Herald Loom companies. Illustrated here is one of the Queens with her Herald loom and Jack Bailey, commentator for the show.

Queens with their looms will be seen all over the country since the program is released to 152 NBC-TV stations and 556 Mutual radio stations. According to surveys, *Queen for a Day* remains the Number One daytime show from 7 A.M. to 5 P.M. on any network.

Handloom prize, from *Handweaver & Craftsman*, Spring 1957.

movement in the mid-1960s, continuing through the 1970s. The Ashford Handicrafts company of New Zealand began making spinning wheels again in 1965, noting in their company history that "the next 20 years (between 1945 and 1965) spinning slumbered like Sleeping Beauty." Their largest market was the US. Back-to-the-landers read books to teach themselves to spin, including *Handspinning* by Allen Fannin published in 1970, *Foxfire 2* by Eliot Wigginton published in 1973, and Paula Simmons's book *Spinning and Weaving with Wool*, published in 1977. By the time

Simmons published her book, she could list 43 makers of spinning wheels. However, in homage to the tradition of the colonial settlers, she also included instructions for making one's own wheel and loom. Rather than reinventing the wheel however, her spinning wheel design could be made using a bicycle rim.

Whether it was a reaction to synthetics, or simple availability, these new spinners mostly were spinning wool. This may have been because some were also raising sheep, and spinning the wool was the logical extension of keeping sheep. Many handspinners were making yarn for hand knitting, a craft also experiencing a revival. Elizabeth Zimmermann, a British emigre to the US, was an adamant defender of wool in the late 1950s, when many knitters had switched to acrylic yarns. She hosted a PBS knitting workshop program. In the 1980s. Allen Fannin, author of *Handspinning*, and his wife Dorothy were technical consultants for an interesting experiment in Vermont attempting to marry the goals of income generation for the poor and teaching of handspinning as an "intermediate technology." The Shelburne Spinners was founded in 1972, with funding from a government grant. The cooperative organization trained spinners, used the renewable resource of Vermont wool and produced wool yarn to sell to knitters and weavers. The organization disbanded in 1978, perhaps after discovering Part F of their purpose, as outlined in their founding document: "To prove that there is or isn't a market for handspun yarn."

The decline of the New England woolen industry and the closure of the Crompton & Knowles Loom Works happened in response to a tsunami of factors. In particular, the development of synthetic fibers was a major challenge to the old industry. However, important changes in lifestyle, the resulting change in demand for wool products, competition in technology, and ultimately competition with imported fabric all contributed to the decline. Nothing could be done to reverse the tide. Wool might be losing its commercial viability, but it was finding its way back to artisan hands, to people who valued the properties of the natural fiber and were consciously choosing it for their hand knit sweaters and handwoven projects.

EPILOGUE
Wool: Necessity to Niche

> "As we look back over this history that which must impress us most strongly is the great variety of the influences which at one time or another have affected the course of the industry."
>
> Chester Wright, *Wool Growing and the Tariff*

Almost a hundred years ago, L. G. Connor, author of *A Brief History of the Sheep Industry in the United States*, presciently wrote, "A new phase in the history of our sheep industry is now developing. This is the return of sheep to the farm. In the future an important and increasing percentage of our meat and wool supply will undoubtedly come from the small farm flock." In his narrative, published before the invention of synthetic fibers, Connor noted that the 7 million sheep in the early 1800s had increased to a high point of 44 million in 1880. How could he have known that there would be only 7 million again in 2003? As of 2016, there are a little over 5 million sheep in the US, out of one billion sheep worldwide. The US Department of Agriculture reported in 2013 on sheep operations: almost all of 795,000 farms (94%) had less than a hundred sheep, and 795 sheep growers (1%) had flocks of over 500 sheep which accounted for 43.4% of the inventory." How, at the height of range herding in the west, did he guess so accurately that 2016 would see small farm flocks again?

Fleece weight is up since colonial times: the average fleece weighs 7.3 pounds as compared with 2 pounds,

Cheviot sheep, York Farm,
Shelburne, Massachusetts, 2006.
PHOTO BY LAUNIE YORK

thanks to careful breeding and the introduction of the Merino. Even accounting for that, wool production is down overall, back to approximately what it was in 1840.

Though the number of sheep in the US today is a quarter of what it was in 1969, per capita consumption of wool hasn't changed much since then. With closets fuller than ever before, the average American buys one to one and a half pounds of wool a year, the equivalent of 2 pairs of socks.

Artifact presentation of Crompton & Knowles loom parts at a mill closing sale, 2004.

Industrial manufacture of cotton and wool textiles, originally the backbone of the New England economy, had all but disappeared by the 1980s. Mill buildings began emptying of their workers in the 1950s. Eventually the buildings were also emptied of the machines, largely being sold for scrap although occasionally they made their way across oceans to live new lives in developing countries. Old cotton and woolen mill buildings were adapted for a variety of reuses—marketplaces, condominiums, offices, other manufacturing enterprises—or they were burned or razed. Of the more than 2,000 at their height, few woolen mills remain. The National Council of Textile Organizations reported in 2007 that 350 textile plants closed between 1997 and 2007, with wool mills experiencing the largest share of the closings. When American textile industry conferences feature talks such as "Automation and Robotics—Present and Future Trends in Product Development" and "Connecting Technical Textiles and Composites," it makes the loom-building technology of Crompton & Knowles and wool weaving almost irrelevant. Wool survived the competitive onslaught of cotton in the nineteenth century, only to be subjected to another interfiber battle with man-made fibers a century later. Now, with a smorgasbord of fiber options in the twenty-first century, wool's share of the market is tiny, being only 1 percent of the fiber used in American mills.

Wool and Politics

Wool has played an important role in wars since the Revolution, and is still considered an element of national security. However, beginning with the Korean War, wool was increasingly being replaced by polyproplylene and acrylic in uniform fabrics. In the Gulf War, this became a problem. In 2006 the Marines banned polyester clothing, due to a substantial rise in burns sustained by US troops wearing "high performance," moisture-wicking undergarments. When exposed to fire, polyester melts and causes more severe damage. Marines were told to wear only cotton undergarments, which would merely burn. Wool, of course, is naturally flame retardant.

Still, the US military consumes 25% of American wool and Congress requires that domestic products be used for the US military when possible (Burlington Industries in North Carolina weaves worsted for traditional military dress uniforms). The Wool Fabric Trust Fund was established in 2000 by Congress to support wool growers, wool manufacturers, and wool garment makers. Its purpose was to offset the loss of sales of domestic wool as a result of duty reductions on imported worsted wool fabric. One arm of the Wool Fabric Trust Fund is the Wool Research, Development, and Promotion Trust Fund, providing grants for research, marketing, product development, and new uses for wool, especially new products for the military. The Wool Fabric Trust Act was reauthorized in 2014, after eight members of Congress wrote to President Barack Obama and Vice President Joe Biden to protest its possible elimination, citing that it was a piece of the "Made in America" push. In a symbolic nod to wool and politics,

US athletes were outfitted in all wool woven peacoats, and knitted sweaters and hats in the closing ceremonies of the 2014 Winter Olympics, made from wool grown in Oregon.

Research has resulted in the developments of fabrics that are shrink proof (machine washable), and this is often a selling point for a wool item. Shrink proofing can be accomplished in different ways, either by coating the scales on the wool shaft with a polymer or resin to keep them from permanently closing when subjected to heat, detergent, and agitation, or by a process of removing the scales altogether with an acid bath. The result, according to manufacturers, is "to develop a breathable, waterproof, stain-repellant, machine washable garment." (Notice that this list did not include warmth? I have my suspicions.)

Industry

As the number of woolen and worsted mills decreased, their nature changed. Instead of the vertical mills of yore which performed all operations from carding and spinning to weaving, today's woolen mills do niche production of single items—blankets, fine woven worsted and knitted woolen garments, knitting yarn, or socks. A handful of specialty mills—Faribault, Woolrich, and Pendleton—still weave blankets. Half of all new Pendleton wool blankets today are sold to Native American tribal members, some being commissioned designs for organizations such as the Tamastslikt Cultural Institute and the American Indian College Fund. The trade blanket tradition begun by seventeenth century traders lives on.

After lying dormant for over sixty years, the Warren mill in Stafford Springs, Connecticut, was bought by a new enterprise, which also purchased the original American Woolen Company trademark. The mill reopened to make high-end worsted fabrics for men's and women's clothing.

Wool socks have proved to be an enduring need, as evidenced by the number of old and new mills making them, proving that a well-turned ankle still gets people's attention. Wigwam and Fox River Mills have both been making socks for over a hundred years, but Smartwool, Darned Tough, Point 6, and Farm to Feet are entirely new enterprises.

A few century-old spinning mills survive: Kraemer Yarns of Nazareth, Pennsylvania, makes knitting yarn, and S&D Spinning Mill in Millbury, Massachusetts, makes, among other things, wool yarn used in the under wrappings of baseballs. Newer spinning mills make wool yarn for knitters and hand weavers; Bartlettyarns, Green Mountain Spinnery, Jagger Spun, Harrisville Designs, and Zeilinger Wool use American-made industrial spinning machinery that may be 100 years old or more. Larger mills process 24,000 to 50,000 pounds of wool a year.

Dozens of smaller mills have opened in the last ten years, using "mini mill" equipment newly engineered for custom spinning of small batches of wool, typically less than 100 pounds of raw fleece, the output of a farm flock.

About fifteen years ago, when there were fewer options for custom spinning in the US, I drove a van full of wool up to MacAusland's, a Canadian mill, to have it spun into yarn for weaving blankets. I stopped at the border, where an agent asked what was in the bags. I answered "raw wool, of no commercial value." He laughed, said "You can say that again!" and waved me on.

S & D Spinning Mill, Millbury, Massachusetts.
PAINTING BY MIKE GRAVES

Marketing and Consumption

As department and mail-order stores were a marketing revolution in the early twentieth century, we are currently experiencing yet another way of buying and selling merchandise, a further fragmenting of the market. What would Paul Cherington make of Amazon, Alibaba, or Etsy?

Dr. Ernest Dichter, speaking at Clemson University's textile marketing forum in 1969, noted that a vast majority of people were "barely aware of what materials their garments are made of." He went on to say that the love people had for cotton or wool had become tired, and called for a new love relationship with natural fibers in "the gourmet age in fibers" with so many options available.

That gourmet fiber age may have arrived in the niche markets in the twenty-first century. Dichter would be pleased to see that it is being rebranded, especially for the benefit of new consumers who grew up after 1969. Companies such as SmartWool, Ibex, Patagonia, Darn Tough, and Ramblers Way (started by the founders of Tom's of Maine) all extoll wool's traditional virtues: breathability, ability to absorb moisture and stay warm, insulation, and odor-absorbing qualities. Excerpts from one company's literature state:

> Chances are, if you're reading this, you've watched the growth of wool. No longer the scratchy insulator limited to a heavy sweater or winter hat, wool is modern, made for progressive living and versatile conditions.
>
> We believe in the interconnectedness of life and that reclaiming an interdependency of American farmland, sheep, and textile manufacturing will benefit local economies, the environment, and our customers.

These companies are Dr. Jaeger's great-grandchildren, following in the footsteps of his "Sanatory Woolen System" back in the 1800s. If he could only hear them extolling the virtues of wool: how it maintains a lower heart rate and limits the rise of body temperature during exercise, and deters lactic acid build up! They also preach of contemporary values: sustainability, renewable resources, the interdependence of agriculture and manufacturing, and the carbon footprint value of buying local. Using wool becomes all at once a lifestyle choice, a political act, and an emotional response to sheep and traditional ways of living.

DIY

Wool is a key ingredient in handspun yarn, scarves, sweaters, socks, shawls, and even felted dryer balls and holiday ornaments. This has generated a proliferation of yarn stores, of small spinneries producing roving for spinners, carded batts for felters, and yarns for knitting and weaving, and of virtual marketplaces such as internet sources of supplies. In our postindustrial society the primordial urge to make things with our hands is spurring a renaissance of interest in knitting, spinning, felting, rug hooking, and weaving. What is it about handmade textiles that moves us to create our own, even after the Industrial Revolution freed us from such labor? Is this urge rooted in needing to measure a day or

Felting class.
Photo by Deb Tewell

Epilogue

Knitting Sheep by Margot Apple.
REPRINTED BY PERMISSION OF HOUGHTON MIFFLIN HARCOURT

a life's progress by miles of yarn spun, yards of cloth woven, inches of sock knitted?

Some handcrafters do it as an artisan business, some incorporate craft into their interpretation of simple living, others use it as an activity to reduce stress. More than a hundred years ago the periodical *Dorcas* opined, "The quiet, even, regular motion of the needles quiets the nerves and tranquilizes the mind, and lets thoughts flow free." Eaton observed in 1937, "Spinning like knitting is restful and relaxing, and for many people the rhythm of the wheel is distinctly soothing." Waldorf schools teach knitting as part of their curriculum, with children learning to knit before they learn to read and write, much as children may have learned 200 years ago.

Looking at sheer numbers, knitting is probably the most common way people interact with wool today. A national survey in 2000 showed 38 million women, or almost one in three, know how to knit or crochet, up from 34.7 million women in 1994. The Craft Yarn Council, representing yarn companies, accessory manufacturers (such as producers of knitting needles), and magazine and book publishers, reported that in the early to mid-2000s, the number of young knitters increased by 150 percent. A 2014 study found that almost half of knitters and crocheters surveyed bought yarn at least once a month. Of course they don't all buy wool, but a good percentage do. Steve Elkins, owner of Webs, a yarn store in

Northampton, Massachusetts, that is a mecca for many knitters, estimates that about 65 percent of the yarn they sell is wool. He is careful to add that their sales are not representative of all retailers—big box stores are a different market—but it is significant that so many knitters choose wool for their hand knit projects.

Knitters of any century appreciate the meditative nature of the activity, but other attractions include the luscious colors and textures of yarns (remember wool takes dye beautifully), the portability of projects, and the social opportunities knitting generates, whether it is sitting with other knitters in real time or with online communities such as Ravelry (with more than 6 million members worldwide as of 2016).

Pastoral scene hooked rug by Margaret Arraj.
USED BY PERMISSION

For the most part, new DIYers may have learned to knit from peers, but usually haven't learned it from their mothers or grandmothers. In addition to the traditional print resources of books and magazines (including the rug hooking periodical the *Wool Street Journal*), aspiring knitters, spinners, felters, and weavers can now teach themselves skills watching YouTube videos, including: how to spin, how to warp a loom, and how to turn a heel on a sock. (Connecting with what was once

Carded vegetable-dyed wool batts. Woolies of Shirkshire Farm, Conway, Massachusetts.

considered a preindustrial skill is now termed "reskilling.") The internet has become a virtual hearth, a social knitwork (ouch) bringing together people to share their passions and knowledge. Social media sites such as Ravelry and Facebook, as well as Yahoo interest groups offer an online community. Shepherds share worming tips and sell fleece; knitters share patterns and troubleshoot each other's projects; spinners watch videos to learn technique; and community members provide detailed instructions for handwashing wool sweaters. Hand producers blog their twenty-first century eclogues to sheep and yarn and post their latest projects. These sites serve as a forum to discuss burning topics such as the advantages and disadvantages of shrink-proof (Superwash™) wool, how it changes the basic structure of the fiber, and whether is it worth the convenience of being able to put it in a washing machine or dryer.

Trying to quantify the number of knitting blogs is like counting the stars in the sky. There are many thousands. One could look at blogging as an online journal, a way of keeping a record, but DIYers also use blogs as a way to solicit online help to solve a knitting problem. Podcasts on all subjects are also available, and there are several dozen podcasts just for knitters.

For those attempting to make a business out of their work, there is Etsy, a vehicle for makers of handcrafted items, primarily women, to sell their work. Etsy is one of the 50 most visited websites in America. Within a half hour, artisans can set themselves up to display their wares, bypassing craft shows, galleries, and other real world markets, and paying only a 3.5 percent commission. The web is another resource for knitters. A search of "handspun wool yarn" yielded 8,766 results, "handspun wool" 10,000 results, and "handspun yarn" 15,000 results. Those looking online can find that 60 yards of handspun wool ranges in price from $10 to $16.

Across the country, old spinning mills have retooled, and small fiber mills have sprung up to prepare wool for handspinning, much the same scale as the carding mills of 200 years ago. Fiber mills can also take a small farm's annual shearing and turn it into yarn.

Taking home a spinning wheel.
Maryland Sheep and Wool show, 2016.

One of the best places to see how wool benefits from this renaissance is to visit a wool festival. Wool festivals are held in at least 18 states: all the New England states, the mid-Atlantic states, and a number of western and midwestern states. Attendance is high at these shows: The New York

Admiring the sheep. Massachusetts Sheep and Woolcraft Fair, 2016.

Lambs ready for the show ring at the Maryland Sheep and Wool show, 2016.

Men at Work Sheep to Shawl team, Rhinebeck, New York Sheep and Wool Show, 2016.

Sheep and Wool Festival had 38,000 attendees in 2016 and the Maryland show has anywhere from 20,000 to 50,000 each year. They are a wonderful showcase for people who value wool. The sheep are the star attraction, and everyone enjoys the sheep shearing and sheep dog demonstrations. The festivals also feature classes, sheep to shawl contests, and "spin ins." The shows are also a prime marketing opportunity for local farms and craft suppliers. I am one of the 250 to 300 vendors at both the Maryland and New York shows and from the vantage point of my booth, I see knitters with basketfuls of multihued wool, spinners and felters dragging wagons of fleece, colorful carded batts, and bags of roving bigger than the attendees themselves. Good natured companions tag along, hoisting spinning wheels, small looms, and shepherd's crooks.

Whither Wool?

Whereas civilization's advance causes fewer people to "experience the frequent deprivation of their basic needs," and as wool originally served to satisfy one of those basic needs, what is its future?

The winds of fashion continue to shift: "Guys are wearing suits again," says Jacob Long, founder of the new American Woolen Company, as rationale for opening a mill to make fine worsteds. A century ago broadcloth cloaks and overcoats were replaced by blue serge suits and lightweight worsted dresses. It may be that wool as an industrial product is making its last stand today as a pair of socks. If all Americans need is two pairs of socks, the new sock manufacturers can supply them, or knitters can buy yarn to knit them.

After a tumultuous century of radical change in our culture, in fibers used in fabrics, and in technology used to produce those fabrics, industrial manufacture of woolens has largely disappeared in the US. However, home manufacture is once again thriving and sheep are back on the farm. Wool may be dismissed as having "no commercial value,"

Two pairs of wool socks.
PHOTO BY AUTHOR

but it has a value beyond measure to knitters, crocheters, spinners, weavers, felters, rughookers, needleworkers, and all the others who choose wool.

Wool consumption in the US over the centuries was born of necessity and availability. In today's world, choosing wool is one of the ingredients in the stew of sustainability. It is at once a political and an emotional decision, arising from affection for sheep and the pastoral life, enjoyment of the fiber, and a preference for renewable resources. Love for wool may once have gotten lost, but now it is found.

Acknowledgments

I have been working on this for a number of years, so I have many people to thank for all the help they provided in a multitude of ways.

First, I need to acknowledge the spinners and weavers, known and unknown, whose work appears in this book. Some textiles have stayed within a family, some are documented in museum collections, some have been created recently enough that we know where they came from. The origins of others are lost in the mists of time, having been found in church rummage sales or as scraps in the bottom of a box.

Secondly, I need to thank all the librarians and curators who so generously assisted me with finding materials. These include, but are not limited to, Liz Jacobson-Carroll of the Buckland Library; Jane Ward and Clare Sheridan of the Museum of American Textile History; Susan Edwards; Jean Hosford of the Complex Weavers Library; Kay Peterson and the staff at the archives of the Smithsonian Museum of American History; the staff at the Hagley Museum Library; the library of the Rhode Island Historical Society and the American Antiquarian Society; Pat and Victor Hilts of the Home Textile Tool Museum; Kara Mitchell of the Jackson County Historical Museum; the Pocumtuck Memorial Hall Museum; Old Sturbridge Village; and Old Deerfield.

Thirdly, I am indebted to all the friends who read drafts of chapters, asked great questions, made wise comments, and generally helped keep the sheep heading in one direction. Thank you Dede Heck, Kate Lamdin, Ken Wells, Fran Rivkin, Jean Hosford, Susan Conover, Marti Taft-Ferguson, and Norma Smayda. Thanks especially to Diane Fagan Affleck, whose knowledge, careful reading, and comments helped it to be a much better book.

Thanks to Catherine Reid, who inspired by example and pointed me towards the American Antiquarian Society.

Acknowledgments

Thanks to my mother, Betty Lou Hart, who taught me to knit and sew, and to my weaving mentors, Helga Reimers and Leonard Brodt.

Thanks to my wonderful sons, Miles Warner, Zakes Warner, and Dan Warner, who provided encouragement and technical support.

Thanks to the Berkshire Taconic Foundation, whose grant allowed me to weave linsey-woolsey and gain a visceral appreciation of what that fabric was.

Thanks to Joanne Semanie, whom I met just at the moment that I was both behind in my commission weaving work and in the final throes of getting all this together. She wound bobbins, knotted fringe, and provided photographs, including a priceless one of the weave room in her family's woolen mill.

Finally, I thank the Turkeyland Cove Foundation and River Road Retreat for providing space and time to do nothing but read, think, and write.

Notes

INTRODUCTION

As late as 1940 a monograph Providence Community Research Center, *Woolen and Worsted Industry Occupational Monograph*, p. 8.

asked in an *American Fabrics* article in 1958 Dichter, "Nine Ways to Vitalize Textile Sales," p. 54.

dating from 9000 BCE Ryder, *Sheep and Man*, p. 3.

The first evidence of wool fabric Ibid., p. 93.

bunting fabric Hayes, *Fleece and the Loom*, p. 62.

after 1920 the market changed rapidly Cherington, *Commercial Problems of the Woolen and Worsted Industry*, p. 42.

COLONIAL PERIOD

In the 51st notebook from 1688 Harte, N.B., "Economics of Clothing," p. 289.

may have included Leicestershires Carman, Heath, and Minto, *Special Report on the History and Present Conditions of the Sheep Industry of the United States*, p. 87.

By 1649, there were said to be 3,000 Ibid., p. 21.

In about 1635, forty Texel sheep American Wool Council, *Story of Wool*, p. 4.

From 1629 to 1640 Thompson, *Mobility and Migration*, p. 202.

This concentrated migration Ibid., p. 184.

the General Court of Massachusetts Ford, *Wool and Manufactures of Wool*, p. 5.

production was 23,400 pounds Mayhew, Aleanor, ed., *Martha's Vineyard*, Dukes County Historical Society, 1966, p. 54.

By 1661, the human population stood Mohanty, *Labor and Laborers of the Loom*, p. 5.

Many were displaced cloth workers Hood, "Industrial Opportunism," p. 136.

In Pennsylvania, more than half Ibid., p. 64.

Colonists wore Langdon, *Everyday Things in American Life*, p. 240.

At the time, linen Bagnall citing Bishop.

Oliver Dickerson writes Baumgarten, *What Clothes Reveal*, p. 76, citing Oliver Dickerson, *Navigation Arts and the American Revolution*.

probate records showed Ouellette, *US Textile Production in Historical Perspective*, p. 78.

This same act forbade As cited in Bagnall, *Textile Industries of the United States*, p. 6.

Another order in 1656 Ibid., p. 6.

Lord Cornbury, Governor Ibid., p. 12.

In 1708 Caleb Heathcote Ibid., p. 12.

George Washington Julie Miller, "George Washington and the Weaving of American History," Library of Congress guest blog post,

March 10, 2015, quoting George Washington, "A Comparison drawn between Manufacturing and importing."

In a diary entry from 1775 Abigail Foote's Journal, June 2, 1775–Sept. 17, 1775, CT Historical Society, as cited in *Home Life in Colonial Days*, p. 253.

Needing 60,000 blankets Papers of George Washington, gwpapers.virginia.edu.

debunking the myth Ouelette, p. 63.

Sometimes the sheep were washed before shearing A sheep washing pool is noted on the map of Catamount Hill in Colrain, Masssachusetts. Catamount Hill was settled in the 1780s, with many families raising sheep. Stone walls crisscross the land, now Massachusetts state forest. Davenport, Elmer, *Puzzle of Catamount Hill*, p. 25.

The second reason US Revenue Commission Special Report #13, *Wool and Manufactures of Wool*, p. 70.

This description of one J. and R. Bronson, *Domestic Manufacturer's Assistant*, Dover Publications, NY, 1977, first published in 1817, p. 42.

"That from persons living on plain diet" William Partridge, *Practical Treatise on Dyeing Woolen, Cotton, and Silk, William Partridge and Son*, NY, originally published 1834, p. 17.

The rule of thumb was Earle, *Home Life in Colonial Days*, p. 198.

3,000 yards were processed annually Ouelette, p. 78.

about a Georgia orphans' house "Itinerant Observations in America" 1745–1746, as cited by Spruill, *Women's Life and Work in the Southern Colonies*.

Most homes had wool blankets Bogdanoff, *Handwoven Textiles of Early New England*, p. 113.

Some had wool curtains *Diary of Samuel Sewall, 1674–1729*, Massachusetts Historical Society, 1878.

Bed curtains Nylander, *Our Own Snug Fireside*, p. 94.

Samuel Lane of Stratham, New Hampshire Lane Family Papers. NH Historical Society as quoted by Brown, *Years of the Life of Samuel Lane*, p. 124.

Planter Robert Beverly Robert Beverly letterbook, Library of Congress, as cited in Baumgartner, *What Clothes Reveal*, p. 135.

NEW NATION

External revenue, import duties, were the source Chernow, *Alexander Hamilton*, p. 339.

one Patriot Pastor Americanus, *The Shepherd's Contemplation*, W. W. Woodward, 1794.

As late as 1791, its production Riello, *Cotton*, p. 203.

Cotton production mushroomed Ibid., p. 203–204.

linseys, satinet The handspun linen warp in linsey-woolsey was replaced by machine spun cotton, and the new mixture known simply as "linsey." Satinet was a woolen-weft-pre-dominant, satin weave cloth structure woven on a cotton warp. It differed from linsey in that it took advantage of the introduction of Merino sheep. Using fine wool in the weft created a softer fabric.

cotton Lowell cloth Just as many British fabrics had been named for their place of origin, "Lowell cloth" identified the plain coarse cotton made on the first cotton power looms in Lowell, Massachusetts.

Clothing was becoming lighter Wass and Fandritch, *Clothing Through American History*, p. 343–346.

The 1790 census Bureau of the Census, *Historical Statistics of the United States*.

Not included in the census Haines and Steckel, *Population History of the United States*, p. 24.

He set up a spinning mill in Canton Bagnall, p. 225.

Mr. Dobson analyzing a sample of cloth Ibid., p. 603.

On Nov. 2, 1830, Joseph Hollingsworth letters, ATHM.

William Morris moved to Ohio Quoted in Erickson, *Invisible Immigrants*, p. 162.

the state of domestic wool was summed up Cole, *Correspondence of Alexander Hamilton*, p. 30.

Connor's report Published in the 1918 Annual Report of the American Historical Association.

In 1802 Colonel David Humphrey Cole, p. 74.

Thousands of acres of land were cleared Wessels, *Reading the Forested Landscape*, p. 58.

In his experience, he allows Bordley, John *Purport of a Letter on Sheep*, Philadelphia 1789; preserved original spelling, where f represents s

Moses Brown wrote Cole, *Industrial and Commercial Correspondence of Alexander Hamilton*, p. 73.

In a letter dated Sept. 6, 1791 Ibid., p. 18.

Another was sent Sept. 27, 1791 Ibid., p. 26.

Alexander King, wrote Ibid., p. 32.

Robert Twiford wrote Ibid., p. 98.

a letter from Sept. 29, 1791 Ibid., p. 100.

Hamilton offers the opinion Ibid., p. 260.

Hamilton also writes Ibid., p. 288.

Finally, in the section on wool, he recommends Ibid., p. 314.

Pattie describes meeting a party Pattie, James, *Personal Narrative*, p. 166.

The tariff rate continued to rise Wright, *Wool Growing and the Tariff*, p. 34.

An ad in the *Pittsfield Sun* Taft, *Introduction of the Woolen Manufacture into the United States*, p. 7.

Elkanah Watson's purchase of Merinos Ibid., p. 8.

By 1810 there were 1,776 carding machines Jeremy, *Transatlantic Industrial Revolution*, p. 126.

In 1825 a spinner using a jack *Wool Technology and the Industrial Revolution*, p. 50.

Cole guesses that half Cole, *American Wool Manufacture*, Vol. 1, p. 207.

Zachariah Allen Ibid., p. 131.

Sarah Snell Bryant Sarah Snell Bryant diaries, Bryant homestead.

Samantha Barrett Steinberg, Sharon, *19th Century Diaries of Samantha and Zeloda Barrett*, p. 117.

An article in the *New Hampshire Gazette* Bagnall, *Textile Industries of the United States*, p. 79.

Putting out peaked around 1820 Dublin, *Transforming Women's Work*, p. 33, quoting Zevin.

Rowland Hazard's correspondence Rowland Hazard Journal, RI Historical Society.

In 1855, the *Providence Journal* reported As quoted in Bagnall, p. 294.

they recorded 1,043,540 yards Tryon, *Household Manufactures in the United States*, p. 147.

Gallatin, Secretary of the Treasury, would report Cited in Wright, *Wool Growing and the Tariff*, p. 20.

An 1830 inventory Ulrich, Laurel, Keynote address to Dublin Seminar, 1999.

produced in New York households Tryon, p. 289.

A census of the industry was taken in 1836 Cole, p. 207.

woolen clothing needs for slaves Hughes, *Thirty Years a Slave*, p. 41.

laws enacted in various states in the south to regulate slave clothing Presentment Grand Jury, Oct. 1822, Charleston, South Carolina.

Harriet Jacobs Jacobs, *Incidents in the Life of a Slave Girl*, from *Norton Anthology of African American Literature*, p. 213.

(Richard Macks) Ed. Rawich, *American Slave: A Composite Autobiography*, Vol. 16, p. 54.

(Clayton Holbert) Ibid., p. 1.

the estate of Abraham Charles Collections and library, Old Sturbridge Village, Sturbridge, Massachusetts.

Elizabeth Hitz concluded Hitz, *Technical and Business Revolution: American Woolens to 1832*, p. 66.

Cole's note Cole, *American Wool Manufacture*, p. 247.

INDUSTRIALIZATION AND TECHNICAL INNOVATIONS

Soon after the invention of the sewing machine Cherington, *The Wool Industry*, p. 193.

availability of cheap cotton especially reconfigured cloth consumption Riello, *Spinning World*, p. 269.

rose to 4 pounds by 1850 Bureau of the Census.

shipped east on the Erie Canal Wentworth, *America's Sheep Trails*, p. 92.

The number of sheep doubled in Ohio Ibid., p. 72.

By 1870, Ohio, Michigan, Indiana, and Illinois Crockett, "Marketing of Wool in the Nineteenth Century," p. 315.

driven to the gold fields to feed hungry miners Wentworth, p. 135.

Churro sheep Connor, p. 136.

Merinos came to Texas *Golden Hoof*, p. 60.

The Valley Farmer, reporting in 1859 As quoted in ibid., p. 78.

Eighty percent were Merino or Merino grade Connor, p. 138.

western islands off the California coast Ibid., p. 204.

the transcontinental railway in 1869 Connor, p. 139.

Robert Maudslay Maudslay, *Texas Sheepman*, p. 125.

The industry doubled between 1859 and 1869 Cole, *American Wool Manufacture*, Vol. 1, p. 378.

the Pacific Mills Ibid., p. 378.

In 1864, 60 million pounds Ibid., p. 377.

a special US Revenue Commission report US Revenue Commission Special Report #13, p. 38.

William McKinley raised protective tariffs Wright, *Wool Growing and the Tariff*, p. 226.

Native Americans were the only group Haines and Steckel, p. 24.

Theo Dodd Wheat, *Blanket Weaving in the Southwest*, p. 41.

in 1871 the US Army *Chasing Rainbows*, p. 12.

From 1868 (with an estimated 5,000 pounds Wheat, Table 2.

until 1875, 60,000 pounds Wheat, p. 52.

Kit Carson Wentworth, p. 546.

Dodd estimated in 1866 Wheat, p. 41.

By 1871, 30,000 sheep Connor, p. 138.

By 1878 Berlant, *Walk in Beauty*, p. 110.

Around 1870, self powered spinning mules *Wool Technology and the Industrial Revolution*, p. 52.

After the combination patent Brewer, *Queen of Inventions*, p. 3.

In 1880, Singer produced Ibid., p. 9.

The Census of 1870 Jenkins, "Western Wool Textile Industry," *Cambridge History of Western Textiles*, p. 773.

Iowa is a good example Connolly, "Home Weaving, Professional Weaving," p. 114.

In 1860, most of the long wool Ibid., p. 132.

the Amana Society Woolen Mill Ibid., p. 110.

The federal census of 1850 Ibid., p. 61.

The US Sanitary Commission Macdonald, *No Idle Hands*, p. 103.

"all wool and a yard wide" *Sterling (Illinois) Gazette*, May 19, 1866.

consumption of wool in the 1850s Langdon, *Everyday Things in American Life*, p. 248.

Ready-to-wear production began *Suiting Everyone*, p. 52.

spread into civilian wear Cole, p. 318.

By the 1860s Ibid., p. 322.

Department stores opened Ley, *Fashion for Everyone*, p. 28.

there was never enough wool grown in the US Wright, p. 75.

Increased consumption may be accounted for by *Ladies Home Journal* no. 7 (15).

GOLDEN AGE OF INDUSTRY

"An increase of nearly 20 per cent in population" US Treasury Department, *Wool and Manufactures of Wool*, p. 8.

Per capita consumption was Langdon, *Everyday Things in American Life*, p. 248.

Sales of consumer goods in general tripled Whitaker, *Service and Style*, p. 1.

the decade of most rapid growth, 1900 through 1910 Ibid., p. 130.

The Corriedale Kupper, *The Golden Hoof*, p. 197.

New American breeds included Wentworth, *America's Sheep Trails*, p. 564.

William Wood Cherington, *Wool Industry*, p. 56.

before the US entered the war in 1917 Roddy, *Mills, Mansions, and Mergers*, p. 78.

In 1917, American Woolen landed Ibid., p. 79.

Shropshire Fraser, *Sheep Husbandry*, p. 90.

The Red Cross *Work of the American Red Cross During the War*, as cited by Macdonald, *No Idle Hands*, p. 199.

The trend towards urbanization Whitaker, p. 1.

African Americans who moved north Florette, *Black Migration*, p. 51.

another great wave of immigration Briggs, *Mass Immigration and the National Interest*, p. 44.

Between 1890 and 1920 Briggs, p. 56.

Manufacturers of clothing had their clothing assembled by contractors Report of the Committee on Manufactures on the Sweating System 52nd Congress 1893.

The New York garment trade produced Crockett, *The Woolen Industry of the Midwest*, p. 102

However, enough data to create sizing standards for women The results were published in USDA Misc. Publication 454, Women's Measurements, 1941.

Cornell University's home economists Tomes, *Gospel of Germs*, p. 145.

"The evil effects upon health" "Standard Underwear of the World," sales brochure, 1911.

"woolen clothing exerts its beneficial influences" Ibid.

golf and cycling suits in 1897 Whitaker, p. 70.

For both men and women, the estimate of 9 pounds *Ladies Home Journal* vol. VII, no. 7, p. 15.

Katherine Fisher Fisher, "Ad-writing and Psychology," as quoted in *Land of Desire*, p. 37.

Its publishers claimed Mrs. John Doe, p. 39.

"A good weaver keeps his belt on a" Cherington, *Commercial Problems of the Woolen and Worsted Industries*, p. 94.

Companies were consolidating Michl, *The Textile Industries*, p. 198.

American Woolen Company continued to acquire mills Roddy, p. 57.

continued to shift from woolens to worsted Cole, as quoted by Crockett, p. 107.

1909 was the high point of worsteds Cherington, *The Wool Industry*, from Tariff Board Report of 1909, p. 103.

invention of an improved combing machine Ibid., quoting US Census data, p. 103.

Census data shows a shift to other methods of power Cole, Vol II, p. 95 graph.

Alice Morse Earl Earle, *Home Life in Colonial Days*, p. 246–247.

Old examples attracted the attention of collectors Ibid., p. 339.

Mrs. Woodrow Wilson Eaton, *Handicrafts of the Southern Highlands*, p. 66.

Frances Goodrich Southern Highlands defined as the mountain regions of Alabama, Georgia, Kentucky, North Carolina, South Carolina, Tennessee, Virginia, and West Virginia.

In 1887, a Commission of Indian Affairs agent reported Berlant, *Walk in Beauty*, p. 116.

The Fred Harvey Company Amsden, *Navaho Weaving*, p. 190.

this example bought in the early 1900s Berlant, p. 142–144.

Max West was tasked to quantify the activity West, "Revival of Handicraft in America," in US Dept. of Labor Bulletin, p. 1596.

WOOL'S GAINS AND LOSSES

"If clothes were worn solely, or chiefly" Cherington, p. 124.

Constance Green Green, *A Case History of the Industrial Revolution*, p. 239.

per capita consumption remained relatively steady International Wool Secretariat, "Textile Fiber Consumption in the United States," p. 8.

"One might perhaps say that the discarding of the steel or whalebone corset" Rees, *St. Michael's: A History of Marks and Spencer*, as quoted in *History of 20th Century Fashions*, p. 137.

A 1950 retrospective *American Fabrics* article "A Changing World," American Fabrics #16, p. 44.

This fueled a home appliance boom Ibid., p. 47.

the number of electric clothes washers Cherington, *Commercial Problems of the Woolen and Worsted Industries*, p. 53.

"Change is in the very air we breathe" Frederick, *Mrs. Consumer*, p. 29.

From 1919 to 1930, expenditures on clothing Michl, *The Textile Industries*, p. 58.

"The winter street costume worn by most women in 1921" Cherington, p. 51.

the location of the market was a moving target Cherington, "Putting American Consumers Under the Microscope, 1924," p. 303.

Per capita consumption of man-made fibers rayon and nylon increased Morris, p. 49, from *World Wool Digest*.

The Jantzen Company US Patent Office, Patent 1,390,135.

In 1910 there was a car for every Cherington, p. 53.

In 1922, a survey of car owners Andries, "Enclosed Car Brings New Art." *Michigan Manufacturer and Financial Record*, p. 6.

"A well executed automobile interior" Ibid.

In a 1940 study, Ernest Dichter wrote Dichter, *Psychological Influence of Women on Car Buying*, p. 79.

the Yale Woolen Mill Carol Andreae Wiegard, *History of the Yale Woolen Mill*, p. 62.

The midpoint of this period, 1935 Garside, *Wool and the Wool Trade*, p. 14.

Consumption of wool dropped from 312.8 million pounds in 1922 Wendzel, *Mill Demand for Wool*, p. 172.

In 1935 the government purchased huge amounts of wool goods Ibid.

Mills rolled out wool fabric for everything from underwear to blankets Mabie, *Neither Wealth nor Poverty*.

Even the American Woolen Company Ibid., p. 129.

"Instead of goods selling themselves" Michl, p. 5.

In 1939, the Census of Manufactures National Association of Wool Manufacturers, *Wool in the United States*, p. 33.

William Wood's successor, Andrew Pierce Roddy, p. 127.

LOSING GROUND: POST–WORLD WAR II

Domestic consumption of man-made fibers increased 73 percent USDA, *Demand for Textile Fibers in the United States*, p. 25.

The Textile Fiber Products Identification Act, effective in 1960 *Kent and Rigel's Handbook of Industrial Chemistry and Biotechnology, Vol. 1*, Kent, 2007, p. 790.

In his address "Tear the Grey Cloak," Ernest Dichter, "Tear the Grey Cloak," p. 1.

Wool was a protector Dichter, Ernest, "Nine Ways to Vitalize Textile Sales," p.57.

Estelle Ellis Estelle Ellis, "Changing Consumer Values," 1994.

a 1958 high school textbook McDermott and Nicholas, *Homemaking for Teenagers*, p. 184–185.

For the second marketing research report USDA, "Women's Attitudes Towards Wool and Other Fibers," p. 4.

They rated wool highly Ibid., p. 6.

Teenage girls were the largest share Ibid., p. 19.

said that they were aware of the fiber content Ibid., p. 21.

In reporting on how they cared for their garments Ibid., p. 35.

Part II of the study Ibid., p. 39.

by 1968 knits represented 25 percent American Fabrics magazine #81.

a 1968 American Fabrics article American Fabrics magazine #81, p. 61.

Ralph Huey Letter from Ralph Huey to Laurence Gross, Director of the American Textile History Museum, collection of Worcester Historical Museum.

The program was phased out in 1995 Accessed at govinfo.library.unt.edu.

As Massachusetts had led the country Koistinen, Confronting Decline, p. 2.

after losing millions of dollars, in 1955 American Woolen Roddy, p. 128.

In 1950 the sheep population Riello, "Counting Sheep," p. 336.

editor Mary Alice Smith asked Mary Alice Smith, Handweaver & Craftsman, Issue 1, 1950.

EPILOGUE

"As we look back over this history" Wright, Wool Growing and the Tariff, p. 319.

L. G. Connor Connor, Brief History of the Sheep Industry in the United States, p. 165.

The US Department of Agriculture reported in 2013 USDA, Farms, Land in Farms, and Livestock Operations, 2013.

Fleece weight is up since colonial times www.sheepusa.com.

350 textile plants closed between 1997 and 2007 NCTO 2007. Trade and Jobs. www.ncto.org/tradejobs/.

wool's share of the market is tiny ASI, 1993.

wool was increasingly being replaced by polypropylene and acrylic www.nap.edu/catalog/12245.html. Changes in the Sheep Industry in the United States, p. 262.

Half of all new Pendleton wool blankets today Colvin, p. 23.

Larger mills process 24,000 Figure supplied by Green Mountain Spinnery.

to 50,000 pounds of wool a year Figure supplied by Bartlett Yarn.

using "mini mill" equipment From Belfast Mini Mill Equipment website

Chances are, if you're reading this Ibex Spring/Summer 2015 catalog.

We believe in the interconnectedness of life www.ramblersway.com/our-story/values-beliefs.

the periodical Dorcas opined Dorcas, March 1884, p. 1.

Eaton observed in 1937 Eaton, Handicrafts of the Southern Highlands, p. 98.

in the early to mid-2000s, the number of young knitters increased Craft Yarn Council, 2002–2004.

The New York Sheep and Wool Festival Figure supplied by Sara Healey, chairman.

the Maryland show Estimate supplied by Judy Sopenski, Maryland Sheep and Wool Festival Committee.

"experience the frequent deprivation" Campbell, "Consuming Goods and the Good of Consuming," Consumer Society in American History, p. 23.

"Guys are wearing suits again" CA Apparel News.

Wool (Mostly) Glossary

Definitions are from Burnham, *Warp and Weft: A Textile Terminology*; Denny, *Fabrics and How to Know Them*; and Montgomery, *Textiles in America, 1650–1870*. Names of textiles often change over time, but I am attempting to define fabrics in the context of the time that I am writing about them. For example, "drugget" later came to be used to describe a coarse rug, but I am using the earlier definition as it referred to apparel fabric.

AMAZON: all wool fabric having a dress face finish which effectively obscures the weave

BAIZE: heavy woolen fabric, felted and napped on both sides

BOMBAZINE: silk warp, wool weft, twill; usually black and used for mourning clothing

BONDED FABRIC: fabric-to fabric construction joined by an adhesive or thin foam

BRILLIANTINE: smooth, wiry material. Cotton warp, lustrous wool weft

BROADCLOTH: wider than 44 inches, plain weave, fine wool, heavily fulled, napped, and sheared

BUNTING: plain weave, loosely woven wool cloth used for making flags

CALIMANCO: glazed woolen cloth

CAMLET (also called Camblet): wool and other fibers, plain weave, used for cloaks and petticoats; patterns made by stamping or water finish

CASSIMERE: medium weight 3 harness twilled woolen cloth

CASSINET: coarser version of satinet: cotton warp, wool weft

CHALLIS: soft lightweight worsted cloth

CHEVIOT: rough, rugged woolen suiting and overcoat cloth, tweed; also a breed of sheep

CHINCHILLA CLOTH: heavy coating with napped surface. Double cloth, may have cotton warp

COFFIN CLOTH: wool cloth with a cotton warp, for the manufacture of shrouds

COVERT: twilled, lightweight wool overcoat cloth made with two shades of a color for warp and weft

DELAINE: fine woolen fabric, worsted weft, cotton warp

Wool (Mostly) Glossary

DOMETT FLANNEL: cotton warp, woolen weft, substituted for linsey-woolsey, later all cotton

DOUBLE KNITS: fabric knitted with two sets of needles forming an interlocking stitch. Less stretchy, won't ravel making it easier to cut and sew

DRUGGET: common cloth, usually plain weave, linen warp and wool weft

DUFFEL: Heavy napped woolen cloth

ECLOGUE: short poem, especially a pastoral dialogue

FEARNAUGHT (ALSO KNOWN AS DREADNAUGHT): thick cloth with a long pile

FINISHING TECHNIQUES: napping, shearing, glazing

FLANNEL: lightweight, soft woolen fabric

FULLING: washing and felting finished cloth

FUSTIAN: cloth made with linen warp, cotton weft

GABARDINE: steep twill weave, very hard wearing

GRENADINE: lightweight dress fabric, later woven in leno

HENRIETTA: lightweight, black silk warp and fine wool weft, used for mourning

HERRINGBONE: a twill in which the diagonal lines are arranged alternately, forming a zigzag

HOMESPUN: in this context, cloth made in the American colonies

HOPSACKING: novelty woolen fabric, rough basket weave

KERSEY: traditional Yorkshire coarse narrow cloth, face finished (one sided) overcoating

KNITTING: a process of creating fabric, using needles to form yarn into a series of interlocking loops

LANOLIN: a fatty substance used in ointments, extracted from wool

LINSEY: plain woven cloth, cotton warp, wool weft

LINSEY-WOOLSEY: coarse cloth made of linen warp and wool weft

MELTON: heavily felted woolen fabric, thicker than broadcloth

MICRON: unit of measurement of wool fiber diameter

NEGRO CLOTH: coarse fabric used for clothing slaves

OVERCOATING: woolen fabric 18 oz. per yard or heavier (broadcloth, melton, kersey)

PLAINS: plain weave, a kind of flannel, used for winter suits

PLAIN WEAVE (ALSO CALLED TABBY): simplest weave structure of weft over and under warp ends

POPLIN: lightweight dress goods, silk warp and woolen weft, resembles bombazine

PRUNELLA: 2/1 warp face worsted twill, often used for women's shoes

RUN: 1600 yards of woolen yarn weighing one pound

SATINET: light mixed cloth, cotton warp, woolen weft

SELVEDGE: woven edge of the cloth

SERGE: men's weight twill fabric, sometimes worsted warp and woolen weft

SHALLOON: cheap twilled worsted, used for dresses and lining

SHIRTING: material woven for shirts; could be cotton, linen, or wool

STAPLE: length of wool fiber

STROUD/TRADECLOTH: heavy woolen cloth, imported and sold in Indian trade especially for blankets

STUFF: lightweight worsted fabric

SUPERWASH™ WOOL: most common treatment for producing machine-washable wool, uses chlorine and Hercosett, a thin polymer resin, to prevent the scales, or cuticles, of the wool fiber from interlocking and causing felting and shrinkage

TAMMY: strong, lightweight worsted

TATTERSALL: checked woolen cloth, used for horse blankets

TWEED: suiting and coating material made of wiry yarn of medium and coarse wool

TWILL: weave structure forming a diagonal line in the cloth

TWIT: a thin place in a piece of yarn, caused by uneven drawing or too much draft in the spinning

UNION CLOTH: cloth woven from a combination of fibers, e.g., wool/cotton

WADMAL (WADMILL): rough woolen felted cloth used for lining horse collars and for rough clothing

WARP: threads set up on the loom

WEAVING: a process of creating fabric by interlacing two or more sets of yarns (warp and weft) at right angles to each other

WEFT: threads used to weave through the warp, also called filling

WHIPCORD: steep twill-like garbardine, long wearing, used for livery coats, riding habits, uniforms

WOOLEN: cloth made of carded woolen fibers, various weights

WORSTED: cloth made of combed long staple wool fibers

ZIBILENE: Heavy coating fabric with long shaggy nap laid in one direction

Bibliography

American Fabrics magazine, 1940–1980. First published by Reporter Publications. New York: Doric Publishing Company.

American Woolen Company Mills. Boston MA: American Woolen Company, 1921.

Amsden, Charles. *Navaho Weaving*. New York: Dover Publications, 1991.

Andries, Henry. "Enclosed Car Brings New Art," *Michigan Manufacturer and Financial Record* (Nov. 11, 1922).

Axtell, James. "The First Consumer Revolution." In *Consumer Society in American History*, edited by Lawrence B. Glickman. Ithaca, NY: Cornell University Press, 1999.

Bagnall, William. *The Textile Industries of the United States*. New York: Augustus Kelley, reprinted 1971.

Baumgarten, Linda. *What Clothes Reveal*. New Haven, CT: Yale University Press, 2012.

Baumgarten, Linda. "Woolens for Slave Clothing," *Ars Textrina* (July 1991).

Benes, Peter, ed. *Textiles in Early New England: Design, Production, and Consumption*. Boston: Boston University Press, 1999.

Benes, Peter, ed. *Textiles in Early New England II: Four Centuries of American Life*. Boston: Boston University Press, 2001.

Benes, Peter, ed. *Women's Work in New England, 1620–1920*. Boston University Press, 2003.

Berlant, Anthony and Mary Kahlenberg. *Walk in Beauty*. Layton, UT: Gibbs-Smith Publisher, 1991.

Bishop, John Leander. *History of American Manufactures 1608–1860*. Philadelphia, PA: E. Young, 1861.

Blaszczyk, Regina. *American Consumer Society, 1865–2005*. Wheeling, IL: Harlan Davidson, 2009.

Blaszczyk, Regina. *Imagining Consumers*. Baltimore, MD: Johns Hopkins University Press, 2002.

Blaszczyk, Regina, ed. *Producing Fashion*. Philadelphia, PA: University of Pennsylvania Press, 2008.

Blum, Herman. *The Loom Has a Brain*. Littleton, NH: Courier Printing Company, 1970.

Bogdonoff, Nancy. *Handwoven Textiles of Early New England*. Harrisburg, PA: Stackpole Books, 1975.

Bordley, John. *Purport of a Letter on Sheep*. Philadelphia: 1789.

Boris, Eileen. *Art and Labor: Ruskin, Morris, and the Craftsman Ideal in America*. Philadelphia, PA: Temple University Press, 1986.

Breweer, Priscilla. *Queen of Inventions: Sewing Machines in American Homes and Factories*. Pawtucket, RI: Slater Mill, 1986.

Bridenbaugh, Carl. *Fat Mutton and Liberty of Conscience*. Providence, RI: Brown University Press, 1974.

Briggs, Vernon. *Mass Industrialization and the National Interest*. Armonk, NY: M.E. Sharpe, 1992.

Brown, Jerald. *Years of the Life of Samuel Lane, 1718–1806*. Hanover, NH: University Press of New England, 2000

Brown, Kathleen. *Foul Bodies: Cleanliness in Early America.* New Haven, CT: Yale University Press, 2009.

Burnham, Dorothy K. *Warp and Weft: A Textile Terminology.* Toronto: Royal Ontario Museum, 1980.

Butterick Publishing Company. *Mrs. John Doe,* 1918.

C&K Loom Pickings, monthly newsletters. Worcester, MA: Crompton & Knowles.

Campbell, Colin. "Consuming Goods and the Goods of Consuming." In *Consumer Society in American History,* edited by Lawrence B. Glickman. Ithaca, NY: Cornell University Press, 1999.

Candee, Richard. "Advertising and Knitting: Cranking out Socks on Contract at Home 1900–1926." In *Proceedings of the Textile History Forum.* Cooperstown, NY, 2003.

Carman, Heath, and Minto, *Sheep Industry of the U.S. Special Report,* USDA, Government Printing Office, 1892.

Cherington, Paul. *Commercial Problems of the Woolen and Worsted Industries.* Washington, DC: Textile Foundation, 1932.

Cherington, Paul. "Putting American Customers Under the Microscope." *Advertising and Selling Fortnightly* 3, 1924.

Cherington, Paul. *The Wool Industry.* Chicago, IL: A.W. Shaw Co., 1926.

Chernow, Ron. *Alexander Hamilton.* New York: Penguin Books, 2005.

Clark, Victor. *History of Manufactures in the United States.* Washington, DC: Carnegie Institution of Washington, 1916.

Clarke, Sarah E. Braddock and Marie O'Mahony. *Techno Textiles 2.* New York: Thames and Hudson, 2006.

Cole, Arthur Harrison. *The American Carpet Manufacture.* Cambridge, MA: Harvard University Press, 1941.

Cole, Arthur Harrison. *American Wool Manufacture,* Vol. 1 and 2. New York: Harper and Row, 1969 (originally published 1926).

Cole, Arthur Harrison, ed. *Industrial and Commercial Correspondence of Alexander Hamilton.* Chicago, IL: A.W. Shaw Co., 1928.

Colvin, Diana. "Trade Blankets Tie Two Cultures." *Oregonian* (Nov. 6, 2003).

Connolly, Deloris. "Home Weaving, Professional Weaving, and Textile Mills in Southeast Iowa, 1833–1870." Master's thesis, Iowa State University, 1982.

Connor, L.G. "A Brief History of the Sheep Industry in the United States," American Historical Association Annual Report 1918.

Cooper, Grace. *The Copp Family Textiles.* Washington, DC: Smithsonian University Press, 1971.

Cooper, Grace. "The Scholfield Wool-carding Machines," *Smithsonian Institution Bulletin* 218, 1959.

Cowan, Ruth. *More Work for Mother.* New York: Basic Books, 1983.

Crockett, Norman. *The Woolen Industry of the Midwest.* Lexington, KY: University Press of Kentucky, 1970.

Crompton, George. *100 Years of Patents.* Worcester, MA: Crompton & Knowles Loom Works, 1937.

Crompton & Knowles Corporation Annual Reports. Worcester, MA.

Crompton & Knowles Loom Works, *C&K Box Loom Census,* Worcester, MA, 1937.

Davenport, Elmer. *Puzzle of Catamount Hill.* Shelburne, MA: Shelburne Historical Society, 1969.

Denny, Grace. *Fabrics and How to Know Them.* Philadelphia: J. B. Lippincott Company, 1928.

Diary of Samuel Sewall. 1674-1729, Massachusetts Historical Society, 1878.

Dichter, Ernest. "Nine Ways to Vitalize Textile Sales," *American Fabrics* No. 43 (Summer 1958).

Dichter, Ernest. *Psychology of Car Buying: Psychological Influence of Women on Car Buying*, 1940.

Dow, G.F. *Arts and Crafts in New England*. New York: Da Capo Press, 1967 (originally published 1927).

Dublin, Thomas. *Transforming Women's Work*. Ithaca, NY: Cornell University Press, 1994.

Earle, Alice Morse. *Home Life in Colonial Days*. New York: Grossett and Dunlap, 1898.

Earle, Alice Morse. *Two Centuries of Costume in America, Vol. 1*, 1903. Reprint, Williamstown, MA: Corner House Publishers, 1974.

Eaton, Allen. *Handicrafts of New England*. New York: Harper and Brothers Publishers, 1949.

Eaton, Allen. *Handicrafts of the Southern Highlands*. New York: Russell Sage Foundation, 1937.

Erickson, Charlotte. *Invisible Immigrants*. Miami, FL: University of Miami Press, 1972.

Ewing, Elizabeth, revised by Alice Mackrell. *History of 20th Century Fashion*. MD: Barnes and Noble Books, 1992.

Ford, Worthington. *Wool and Manufactures of Wool*. Washington, DC: Government Printing Office, 1894.

Foster, Helen. *New Raiments of Self*. Oxford, UK: Berg, 1997.

Frederick, Christine. *Selling Mrs. Consumer*. New York: Business Bourse, 1929.

Gehret, Ellen. *Rural Pennsylvania Clothing*. York, PA: Liberty Cap Books, 1976.

Goodrich, Frances Louisa. *Mountain Homespun*. Knoxville, TN: University of Tennessee Press, 1989.

Green, Constance. *A Case History of the Industrial Revolution*. New Haven, CT: Yale University Press, 1939.

Green, Nancy. *Ready to Wear and Ready to Work*. Durham, NC: Duke University Press, 1997.

Grizzard, Frank. "Supply Problems Plagued the Continental Army from the Start." http://gwpapers.virginia.edu/history/articles/Supply-problems-plagued-th-continental-army-from-the-start.

Haines, Michael and Richard Steckel. *Population History of the United States*, Cambridge, MA: Cambridge University Press, 2000.

Handley, Susannah. *Nylon: The Manmade Fashion Revolution*. London, England: Bloomsbury Press, 1999.

Harte, N.B. "Economics of Clothing in the Late Seventeenth Century." *Textile History* (Autumn 1991).

Hayes. "Fleece and the Loom." Address before the National Association of Wool Manufacturers, 1865 Henri, Florette. *Black Migration: Movement North, 1900-1920*. New York: Anchor Press, 1975.

Hazard, R.G. Papers, manuscript. Rhode Island Historical Society.

Hilts, Patricia. "Preserving the Textile Records of the Reedsburg Woolen Mill of Reedsburg, Wisconsin: An Exercise in Curatorship." Master's thesis, University of Wisconsin, 1976.

Hilts, Victor, and Patricia Hilts."Not for Pioneers Only: the Story of Wisconsin's Spinning Wheels," *Magazine of State Historical Society of Wisconsin* (Fall 1982).

Hindle, Brooke and Steven Lubar. *Engines of Change: The American Industrial Revolution, 1790-1860*. Washington, DC: Smithsonian. Institution Press, 1986.

Hitz, Elizabeth. "Technical and Business Revolution: American Woolens to 1832." PhD dissertation, NY University, 1978.

Hood, Adrienne. *The Weaver's Craft.* Philadelphia, PA: University of PA Press, 2003.

Hughes, Louis. *Thirty Years a Slave.* Electronic edition available at: www.docsouth.unc.edu.

Jenkins, D., ed. *Cambridge History of Western Textiles*, Vol. 1 and 2. Cambridge, UK: Cambridge University Press, 2003.

Jenkins, D.T. and Ponting, Kenneth G. *The British Wool Textile Industry 1770-1914.* Aldershot, England: Scholar Press, 1982.

Jeremy, David. *Transatlantic Industrial Revolution: The Diffusion of Textile Technologies between Britain and America, 1790-1830s.* Cambridge, MA: MIT Press, 1981.

Kidwell, Claudia and Margaret Christman. *Suiting Everyone.* Washington, DC: Smithsonian University Press, 1974.

Koistinen, David. *Confronting Decline.* Gainesville, FL: University Press of Florida, 2013.

Kulik, Gary. *New England Mill Village 1790-1860.* Cambridge, MA: MIT Press, 1982.

Kupper, Winifred. *The Golden Hoof.* New York: Knopf, 1945.

Langdon, William Chauncy. *Everyday Things in American Life.* New York: Charles Scribner's Sons, 1941.

Leach, William. *Land of Desire.* New York: Random House, 1993.

Leavitt, Thomas, ed. *The Hollingsworth Letters: Technical Change in the Textile Industry, 1826-1837.* Cambridge, MA: Society for the History of Technology and the M.I.T. Press, 1969.

Leblanc. *Location of Manufacturing in New England in the 19th Century.* Hanover, NH: Geography Publications at Dartmouth, 1969.

Ley, Sandra. *Fashion for Everyone: The Story of Ready-to-Wear.* New York: Charles Scribner's Sons, 1975.

Licht, Walter. *Industrializing America.* Baltimore, MD: Johns Hopkins University Press, 1995.

Linton, George. *Applied Basic Textiles.* New York: Duell, Sloan and Pearce, 1966.

Little, Frances. *Early American Textiles.* New York: Century Company, 1931.

Macdonald, Anne. *No Idle Hands: The Social History of American Knitting.* New York: Ballantine Books, 1988.

Maudslay, Robert. *Texas Sheepman.* Austin, TX: University of Texas Press, 1951.

Mayhew, Aleanor, ed. *Martha's Vineyard.* Edgartown, MA: Dukes County Historical Society, 1966.

Mazur, Paul. *The Standards We Raise.* New York: Harper Brothers, 1953.

Merrimack Valley Textile Museum. *Wool Technology and the Industrial Revolution.* Exhibit Catalog, 1965.

Michl, H.E. *The Textile Industries.* Washington, DC: Textile Foundation, 1938.

Mohanty, Gail. *Labor and Laborers of the Loom.* New York: Routledge, 2006.

Montgomery, Florence. *Textiles in America.* New York: Norton Books, 2003.

Morris, James. *Woolen and Worsted Manufacturing in the Southern Piedmont.* Columbia, SC: University of South Carolina Press, 1952.

National Association of Wool Manufacturers. "Textile Education Amongst the Puritans." New York, 1911.

National Association of Wool Manufacturers. *Wool in the United States.* New York, 1947.

National Association of Wool Manufacturers Bulletins. Wakefield, MA: Murray Printing Company.

National Wool Growers Association. "Men, Sheep, and 100 Years." Salt Lake City, Utah, 1965.

Nylander, Jane. *Our Own Snug Fireside*. New Haven, CT: Yale University Press, 1994.

Ouellette, Susan. *US Textile Production in Historical Perspective*. New York: Routledge, 2007.

Pattie, James and Thomas Flint, ed. *Pattie's Personal Narrative of a Voyage to the Pacific and in Mexico, June 20, 1824-August 30, 1830*. Carlisle, MA: Applewood Books, originally published 1831.

Pendleton Woolen Mills. *The Romantic Story of Man and Sheep*. 1962.

Providence Community Research Center. *Woolen and Worsted Industry Occupational Monograph*. Providence, RI: 1939.

Rawich, George P., ed. *American Slave: A Composite Autobiography*. Westport, CT: Greenwood Publishing Company, 1972.

Riello, Giorgio. *Cotton: The Fabric that Made the Modern World*. Cambridge, UK: Cambridge University Press, 2013.

Riello, Giorgio. "Counting Sheep." In *Wool: Products and Markets,* ed. by Fontana and Gayot. Padova, Italy: CLEUP, 2004.

Ripley, Fuller. *Fabric of Troy*. Troy, NH: Troy Mills 1986.

Roddy, Edward. *Mills, Mansions, and Mergers*. Lowell, MA: American Textile History Museum, 1982.

Rogers, Grace. "The Scholfield Wool-Carding Machines." Bulletin 218: Contributions from the Museum of History and Technology, Smithsonian Institution, 1959.

Ryder, M.L. *Sheep and Man*. London, England: Duckworth and Co., 1983.

Schlesinger, Arthur. *Rise of the City. History of American Life*, Vol. 10, New York: MacMillan Co., 1933.

Shaw, Madelyn. "Slave Cloth and Clothing Slaves," *MESDA Journal*, available online:www.madelynshaw.com.

Soyer, Daniel, ed. *A Coat of Many Colors*. New York: Fordham University Press, 2005.

Steinberg, Sharon. "19th Century Diaries of Samantha and Zeloda Barrett." In Proceedings of the Textile History Forum. Cooperstown, NY, 2003.

Stoll, Stephen. *Larding the Lean Earth*. New York: Hill and Wang, 2002.

Strasser, Susan. *Waste and Want*. New York: Metropolitan Books, 1999.

Strawn, Susan. *Knitting America*. Minneapolis, MN: Voyageur Press, 2007.

Taft, Royal. *The Introduction of the Woolen Manufacture into the United States*. Cambridge, 1871.

Thompson, Roger. *Mobility and Migration: East Anglican Founders of New England 1629-1640*. Amherst, MA: University of Masssachusetts Press, 1994.

Tomes, Nancy. *Gospel of Germs*. Boston, MA: Harvard University Press, 1999.

Tryon, Rollo. *Household Manufactures in the United States 1640-1860*. Chicago, IL: University of Chicago Press, 1917.

United States Department of Agriculture. "Fabrics and Fibers for Passenger Cars." Marketing Research Report No. 152, 1957.

———. "Women's Attitude Towards Wool and Other Fibers." Marketing Research Report No. 153, 1957.

———. "Teenage Girls Discuss Their Wardrobes." Marketing Research Report No. 155, 1957.

———. "The Demand for Textile Fibers in the United States." USDA Technical Bulletin No. 1301, 1963.

———. "Consumers Concepts of Fabric." Marketing Research Report No. 338, 1959.

———. "Farms, Land in Farms, and Livestock Operations 2012 Summary." February 2013. http://usda.mannlib.cornell.edu/usda/nass/FarmLandIn//2010s/2013/FarmLandIn-02-19-2013.pdf.

United States Department of Commerce, Bureau of the Census. "Population of States and Counties of the US: 1790–1990." March 1996.

United States Revenue Commission. "Wool and Manufactures," Special Report #13. Washington, DC: Government Printing Office, 1916.

United States Treasury Department. *Wool and Manufactures of Wool*. Washington, DC: Government Printing Office, 1894.

Wass, Ann Buermann and Michelle Webb Fandritch. *Clothing Through American History 1786-1860*. Santa Barbara, CA: Greenwood, 2010.

Wendzel, Julius. *Mill Demand for Wool and Inter-Textile Competition. Review of Economics and Statistics*, Vol. 18, No. 4 (Nov. 1936).

Wessels, Tom. *Reading the Forested Landscape*. Woodstock, VT: Countryman Press, 1997.

West, Max. "Revival of Handicraft in America." US Dept. of Labor Bulletin #55 (Nov. 1904).

Wheat, Jon Ben. *Blanket Weaving in the Southwest*. Tucson, AZ: University of Arizona Press, 2003.

Whitaker, Jan. *Service and Style*. New York: St. Martins Press, 2006.

Wiegand, Carol. "History of the Yale Woolen Mill, 1881–1963." Master's thesis, Michigan State University, 1982.

Wills, Kerry. *The Close-Knit Circle*. Westport, CT: Praeger Publishers, 2007.

Wool Bureau. *Story of Wool*. Denver, CO: Wool Education Center, 1968.

Wright, Chester. *Wool Growing and the Tariff*. Boston, MA: Houghton Mifflin, 1910.

Index

American Fabrics magazine, 115, 134–136, 140, 143, 144, 145, 148, 149
American Wool Council, 129, 145, 147
American Woolen Company, 29, 106, 107, 128, 129, 153, 161, 170
Arkwright, Richard, 32, 34, 43, 52, 57
Arts and Crafts Movement, 108
Atwater, Mary Meigs, 109–110, 154
Automobile upholstery, 107, 115, 123, 124, 126, 138
Bathing suits, 13, 99–100, 122
Beacon Blanket Company, 128
Bed curtains, 38, 39, 69
Berea College, 111
Bighorn sheep, 9
Blankets, 9, 11, 13, 15, 25, 28, 38, 39, 54, 72, 74, 76, 79, 84, 85, 93, 101, 103, 107, 111, 128, 146, 152, 161
British woolens, 24, 25, 26, 28
Broadcloth, 15, 16, 18, 19, 20, 21, 24, 37, 49, 57, 63, 70, 92, 107, 125, 170, 182
Buy America Act, 127
Carding: hand, 27, 31, 32; machine, 45, 56, 57, 58, 59, 62, 64, 66, 86, 168
Census, population, 25, 42, 44, 74
Census of Manufactures, 10, 65, 66, 86, 108, 125, 129, 130, 131
Central heating, 6, 92, 114, 118
Chatham Blanket Company, 103, 125, 126
Cherington, Paul, 13, 113, 118
Cloak, 13, 15, 19, 38, 87, 100, 170, 182
Cole, Arthur, 16, 51, 59, 70
Cotton fiber, 15, 42, 43, 86, 97, 120, 135, 137, 143
Cotton yarn, 44, 45, 51, 58, 59, 63, 64, 78
Cotton loom, 47, 60, 66, 72, 80
Coverlets, 109, 110
Coxe, Tench, 42, 65, 66

Crompton, George, 105
Crompton, Samuel, 32, 34
Crompton, William, 34, 60, 72, 80
Crompton and Knowles Loom Works, 12, 80, 105, 106, 133, 150, 151, 156
Dacron, 15, 134, 135
Department stores, 87, 92, 104, 107, 163
Dichter, Ernest, 6, 118, 123, 135, 136, 163
DIY, 166, 167
Dobby loom, 60
Dr. Jaeger's Sanatory Woolen System, 98, 163
Domestic Manufacturer's Assistant, 62
Double knits, 148
Dupont, 48, 49, 120, 121, 134
Eclogue, 8, 167, 183
Electrification, 96, 114, 115
Ellis, Estelle, 137
Etsy, 163, 167
Faribault Mill, 85, 107, 161
Frederick, Christine, 117
Fulling, 20, 29, 32, 37, 56, 183
Goodrich, Frances, 110
Grey's Raid, 22
Hamilton, Alexander, 41, 51, 54, 65
Handloom, 36, 45, 64, 66, 133, 154, 155
Handweaving, 20, 66, 86, 110, 133, 154
Hargreaves, James, 32, 34
Hartford Manufacturing Company, 50, 63
Harvey, Fred, 103, 111, 112
Hindman Settlement School, 110
Hollingsworth, Joseph, 45–47
Immigrants, 10, 23, 44, 45, 72, 95, 96
Imported cloth, 21, 25, 27, 28, 38, 40, 49, 54, 55, 67, 69, 70, 160
Jantzen, 100, 122
Jarvis, William, 49
King, Gregory, 18
Knitting, 11, 37, 87, 105, 143, 156, 165, 167, 183

Knitting machine, 83, 95, 99, 100, 149
Knowles, Lucius, 81, 105
Ladies Home Journal, 104, 116, 118, 125
Lanolin, 30, 37, 183
Linen, 15, 18, 24, 25, 26, 27, 43, 67, 121
Louisiana Purchase, 42
Man-made fibers, 6, 15, 114, 120, 127, 131, 132–135, 147, 151, 154, 160
Marketing, 64, 97, 104, 114, 117, 118, 121, 136, 137, 138, 160, 163, 170
Martha's Vineyard, 21, 22
Mayflower, 20, 21
National Association of Wool Manufacturers, 118, 135, 152, 153
National Wool Act, 152
National Wool Growers Association, 78
Nantucket, 9
Native Americans, 10, 25, 44, 47, 54, 74, 79, 161
Navajo, 54, 79, 111, 112
Orlon, 15, 134, 135, 141, 145
Patents, 12, 55, 81, 83
Pendleton Woolens, 5, 103, 161
Penland, 110
Per capita consumption, 13, 15, 40, 66, 70, 72, 86, 90, 92, 101, 115, 133, 159
Population: census, 25, 42, 44, 74, 75; colonial, 25, 26; slave, 44, 67; Native American, 44, 79
Polyester, 134, 146, 160
power loom, 56, 59, 64, 65, 66, 72, 84
Power: water, 11, 57, 84, 108; steam, 108
Properties of wool, 7, 8, 84, 135, 156
Railroad, 74, 112
Ravelry, 166, 167
Rayon, 12, 14, 114, 117, 119, 120, 122, 126, 130, 134, 139, 140, 154
Ready-to-wear, 87, 92, 96, 97,

101, 105, 128
Report on the Subject of Manufactures, 41, 47, 51, 63, 65
Satinet, 44, 45, 47, 58, 59, 62, 66, 67, 87, 184
Scholfield brothers, 45, 56, 57, 64
Scouring, 30, 36, 50
Sewing machine invention, 11, 72, 78, 83
Shearing, 29, 30, 49, 78
Sheep breeds: Corriedale, 48, 93; Merino, 10, 18, 48, 49, 50, 57, 66, 75, 78, 79; Shropshire, 79, 93, 95; Tunis, 48; Wiltshire, 20, 21
Sheep washing, 30
Singer, 72, 80, 83
Skirting, 29
Slater, Samuel, 45, 51, 57
Slave cloth, 24, 27, 28, 38, 39, 40, 64, 65, 67–69
Spinning: hand, 32–33, 34, 58, 59, 63, 66, 86, 108, 154, 155–156, 165; jack, 33, 58, 81; jenny, 33, 34, 45, 58, 62, 64
Sulzer loom, 131, 149, 151
Superwash wool, 167, 184
Tariffs, 11, 55, 78, 79, 92, 93, 94, 137–138, 152, 157
United States Department of Agriculture, 10, 94, 137, 138
US military, 11, 51, 101, 130, 133, 160
USDA, *see* United States Department of Agriculture
Wars: Civil, 11, 48, 73, 75, 76–78, 85, 86, 87; Revolutionary, 50; 1812, 45, 49, 54; World War I, 93, 107, 109, 110, 117, 122; World War II, 9, 115, 121, 128, 130
Washington, George, 27, 28, 48, 50, 63
Wash and wear, 133, 147

Washing machine, 115, 116, 134, 167
Wentworth, Edward, 16
West, Max, 112
Wilson, Woodrow, 94, 109, 110
Wool Products Labeling Act, 127
Woolen cloth, 12, 18, 24, 27, 29, 32, 38, 41, 50, 55, 59, 62, 63, 64, 65, 66, 67, 69, 78, 80, 84, 85, 86, 92, 107, 125
Woolrich, 85, 146, 161
Worsted cloth, 12, 13, 18, 67, 72, 85, 92, 93, 96, 101, 108, 129, 160, 161, 170
Zimmermann, Elisabeth, 156